PRIVATE PILOT'S *Dictionary & Handbook*

PRIVATE PILOT'S *Dictionary & Handbook*

Second Edition, Revised and Expanded

by KIRK POLKING

ARCO PUBLISHING, INC.
NEW YORK

Second Edition, First Printing, 1986
Published by Arco Publishing, Inc.
215 Park Avenue South, New York, N.Y. 10003

Library of Congress Cataloging-in-Publication Data

Polking, Kirk,
The private pilot's dictionary and handbook.

1. Aeronautics—Dictionaries. 2. Private flying—Dictionaries. I. Title.
TL509.P64 1985 629.132′5′0321 85-18700
ISBN 0-668-05920-6

Printed in the United States of America

10 9 8 7 6 5 4 3 2 1

The author appreciatively acknowledges the advice and suggestions of the following: Federal Aviation Administration personnel in Washington D.C. and at Greater Cincinnati and Lunken Airports in Cincinnati; Cardinal Air Training, the Lunken Airport Administrator, and the Aircraft Owners and Pilots Association in Frederick, Maryland.

Preface to the Second Edition

The *Private Pilot's Dictionary and Handbook* is designed for the student pilot or VFR (visual flight rules) pilot who'd like a convenient reference dictionary to the most commonly used terms, operation procedures, and rules for VFR flight.

It is distilled from the *Airman's Information Manual* and appropriate Federal Aviation Regulations, as of July 1984.

It is designed as a supplement to any ground school course or flight training undertaken by the student and private pilot, and concentrates on single-engine aircraft.

It includes only those IFR (instrument flight rules) terms or information which the VFR pilot might see listed on charts or hear in radio communications but not be required himself to use.

It includes explanations of the abbreviations used in FAA publications and illustrations where they enhance the definitions.

This book is printed in black and white; therefore where colors, such as magenta or blue, were used in charts, phrases such as "printed in magenta" will appear under the appropriate symbols.

This new edition reflects all the changes that have taken place since the book was originally published in 1974. It incorporates new regulations, such as the biennial flight review requirement, and includes the terms most frequently used in pilot/air traffic controller communications, as well as the most common weather abbreviations.

A

A in the phonetic alphabet is Alfa (pronounced al-fah).

A—Symbol for arctic air mass; also weather symbol for hail.

AAS—(*See* AIRPORT ADVISORY SERVICE.)

AAWS—automatic aviation weather service.

Abeam—An aircraft is "abeam" a fix, point or object when that fix, point or object is approximately 90 degrees to the right or left of the aircraft track. Abeam indicates a general position rather than a precise point.

Abort—To terminate a preplanned aircraft maneuver, e.g., an aborted takeoff.

absolute altitude—(*See* ALTITUDE.)

A/C—approach control.

acceleration error—magnetic compass error that results from the tendency of the compass to point down as well as north. This error is most apparent on flight headings of east or west. If the aircraft is heading east or west, an increase in airspeed causes the compass to indicate a turn toward north, while a decrease causes it to indicate a turn toward south. (When the aircraft is on a north or south heading, no compass error is apparent while changing airspeed.) It's best, therefore, to read the compass only when the aircraft is flying straight and level at a constant airspeed.

accident cause factors—

1. The 10 most frequent cause factors for general aviation accidents that involve the pilot in command are:

   a. inadequate preflight preparation and/or planning;

accident cause factors *continued*

b. failure to obtain and/or maintain flying speed;
c. failure to maintain direction control;
d. improper level off;
e. failure to see and avoid objects or obstructions;
f. mismanagement of fuel;
g. improper in-flight decisions or planning;
h. misjudgment of distance and speed;
i. selection of unsuitable terrain;
j. improper operation of flight controls.

2. For information on accident prevention activities contact your nearest General Aviation or Flight Standards District Office.

3. *Alertness*—Be alert at all times, especially when the weather is good. Most pilots pay attention to business when they are operating in full IFR weather conditions, but strangely, air collisions almost invariably have occurred under ideal weather conditions. Unlimited visibility appears to encourage a sense of security which is not at all justified. Considerable information of value may be obtained by listening to advisories being issued in the terminal area, even though controller workload may prevent a pilot from obtaining individual service.

4. *Giving way*—If you think another aircraft is too close to you, give way instead of waiting for the other pilot to respect the right-of-way to which you may be entitled. It is a lot safer to pursue the right-of-way angle after you have completed your flight.

accidents, procedure for reporting—(*See* AIRCRAFT ACCIDENT AND INCIDENT REPORTING.)

accuracy landings—Landings to a marked spot on a runway, also called spot or precision landings, require skill on the part of the pilot in knowing his aircraft's performance, current wind conditions, etc.

acft—aircraft.

acknowledge—A phrase used in radio communications meaning "let me know that you have received and understood this message."

acrobatic flight—A maneuver involving an abrupt change in an aircraft's attitude, an abnormal attitude, or abnormal acceleration, not necessary for normal flight.

acrobatics—Federal Aviation Regulations prohibit acrobatics in an aircraft flying over a congested area of a city, town, or settlement, or over an open air assembly of persons; within a control zone or federal airway, below 1,500 feet above surface, or when visibility is under 3 miles.

actv—Active.

ADF—Automatic direction finder

adiabatic lapse rate—The rate of change of temperature with height. Normally, the temperature decreases with increasing altitude approximately 3.5 degrees Fahrenheit per 1,000 feet. But this is only an average, and it varies with the type of air (dry, moist) ascending from the earth, with altitude, and with other conditions.

ADIZ—Air defense identification zone.

admin—Administration.

advection fog—Fog created by the horizontal movement of warm, moist air over cooler land or water. Mostly occurs during the night, when the earth is losing heat by radiation.

advise intentions—A phrase used in radio communications meaning "Tell me what you plan to do."

advisory service—Advice and information provided by a facility to assist pilots in the safe conduct of flight and aircraft movement.

aerobatics—The abrupt change in an aircraft attitude in excess of 60 degrees of bank and 30 degrees of pitch.

aeronautical advisory stations (UNICOM)—The radio frequencies 123.0, 122.7 and 122.8 MHz are assigned to airports not served by a control tower. The pilot can obtain information on runway and wind conditions, etc., from the airport operator, but there is no air traffic control. Frequency 122.950 MHz is assigned to airports *with* a control tower, and although the pilot would get runway and wind information from the tower, he could use the UNICOM frequency to ask for a phone call to be made or a cab to be ordered to the airport. The frequency reserved for private airports (not open to the public) and air-to-air communications is 122.750.

Airports having Control Towers (Airport Traffic Areas) are shown in blue, all others in magenta.
Consult Airport/Facility Directory (A/FD) for details involving airport lighting, navigation aids, and services.

AIRPORTS

Other than hard-surfaced runways

Hard-surfaced runways 1500 ft. to 8000 ft. in length

Hard-surfaced runways greater than 8000 ft.

Seaplane Base (SPB)

All recognizable hard-surfaced runways, including those closed, are shown for visual identification.

ADDITIONAL AIRPORT INFORMATION

Private "(Pvt)" – Non-public use having emergency or landmark value.

Military – Other than hard-surfaced. All military airports are identified by abbreviations AFB, NAS, AAF, etc. For complete airport information consult DOD FLIP.

Heliport – Selected

Abandoned – paved having landmark value.

Unverified

Services – fuel available and field tended during normal working hours depicted by use of ticks around basic airport symbol. Consult A/FD for service availability at airports with runways greater than 8000 ft.

☆ Rotating light in operation Sunset to Sunrise.

AIRPORT DATA

Box indicates Special Traffic Area (See FAR 93) → NAME
FSS ← Indicates Flight Service Station on field.
NAME CT – 118.3★
ATIS 124.9
03 L 92 122.95 ← UNICOM
VFR Advsy 125.3
Airport of entry

FSS – Flight Service Station
CT – 118.3 – Control Tower (CT) – primary frequency
★ – Star indicates operation part time. See tower frequencies tabulation for hours of operation.
ATIS 124.9 – Automatic Terminal Information Service
UNICOM – Aeronautical advisory station
VFR Advsy – VFR Advisory Service shown where ATIS not available and frequency is other than primary CT frequency.
03 – Elevation in feet
L – Lighting in operation Sunset to Sunrise
*L – Lighting available on request, part-time lighting, or pilot-controlled lighting.
92 – Length of longest runway in hundreds of feet; usable length may be less.
S – Normally sheltered take-off area (SPB)
When facility or information is lacking, the respective character is replaced by a dash. All lighting codes refer to runway lights. Lighted runway may not be the longest or lighted full length.
All times are local.
NFCT – Non Federal Control Tower

Figure A-1. Airport symbols and data.

aeronautical charts—Give pilots geographical information, airport data, obstacles and radio-frequency details. They come in several forms: *Sectional* charts with a scale of 1 inch equaling approximately 6.86 nautical miles are used in most private flying (*See* Figure A-2). (There are enlargements of major local terminals printed right on the sectionals the way downtown city enlargements are printed on state highway maps.) *World* aeronautical charts have a scale of 1 inch

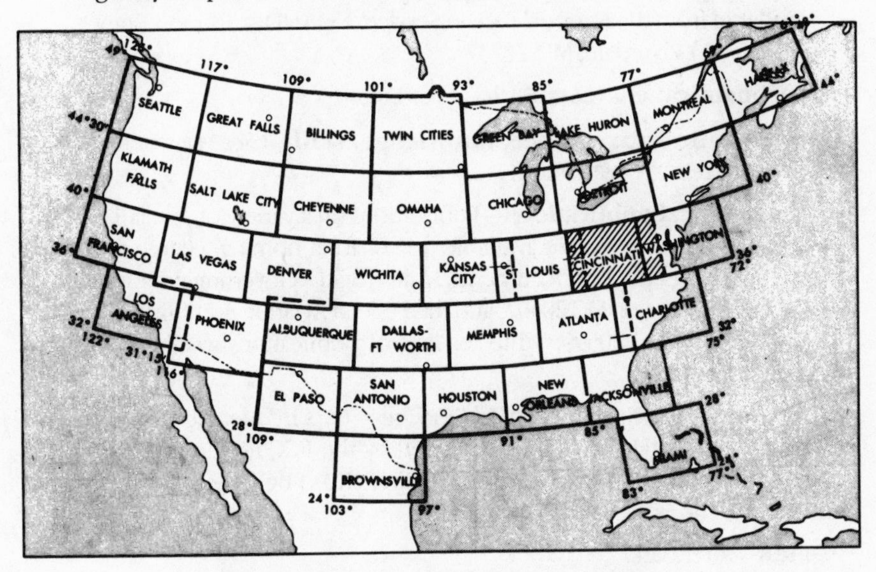

SECTIONAL AERONAUTICAL CHART
SCALE 1:500,000

Lambert Conformal Conic Projection Standard Parallels 33°20′ and 38°40′
Topographic data corrected to May 1984 R1

32 ND EDITION *August 2, 1984*
Includes airspace amendments effective *July 5, 1984*
and all other aeronautical data received by June 14, 1984
Consult appropriate NOTAMs and Flight Information
Publications for supplemental data and current information.
This chart will become *OBSOLETE FOR USE IN NAVIGATION* upon publication of
the next edition scheduled for *JANUARY 17, 1985*
PUBLISHED IN ACCORDANCE WITH INTER-AGENCY AIR CARTOGRAPHIC COMMITTEE
SPECIFICATIONS AND AGREEMENTS. APPROVED BY:
DEPARTMENT OF DEFENSE ∗ FEDERAL AVIATION ADMINISTRATION ∗ DEPARTMENT OF COMMERCE

Figure A-2. Sectional aeronautical charts.

aeronautical charts *continued*

equaling approximately 13.7 nautical miles and are used for long flights at high speeds. *Operational navigation* charts are the same as the world aeronautical charts and show essentially the same information, but they also include elevations shown in shaded relief as well as contour. VFR terminal area charts are now also available for high traffic airports with a scale of 1 inch equaling 3.43 nautical miles. Charts are available from the National Ocean Service, NOAA Distribution Branch (N/CG 33) Riverdale, MD 20737.

aeronautical chart symbols—(*See* Figure A-3.)

aeronautical experience, knowledge, skill—(*See* PREREQUISITES FOR PRIVATE PILOT CERTIFICATE.)

aeronautical publications—Publications on a variety of specialized aeronautical subjects are available free or for a nominal charge from the Superintendent of Documents, United States Government Printing Office, Washington DC 20402, and the Federal Aviation Administration, Washington, DC 20590. Write for lists of publications available and their costs.

Aeronautical Utility Mobile Service—Two frequencies, 122.775 and 122.850 MHz, have been assigned by the FCC to allow pilots to communicate with ground service units such as FBOs, fuel trucks, and avionics shops.

afdk—After dark.

affirmative—A phrase used in radio communications, meaning "yes."

agl—Above ground level.

Agriculture Department (USDA) restrictions—The importation of plants, plant products, all birds, certain animals, meats, and meat and animal products is regulated by the U.S. Department of Agriculture to prevent the introduction of plant and animal pests and diseases. Agricultural items should not be brought to the United States in lunches or otherwise unless you are informed in advance by Agriculture inspectors of the Animal and Plant Health Inspection Service (APHIS) or Customs officers that such items are admissible.

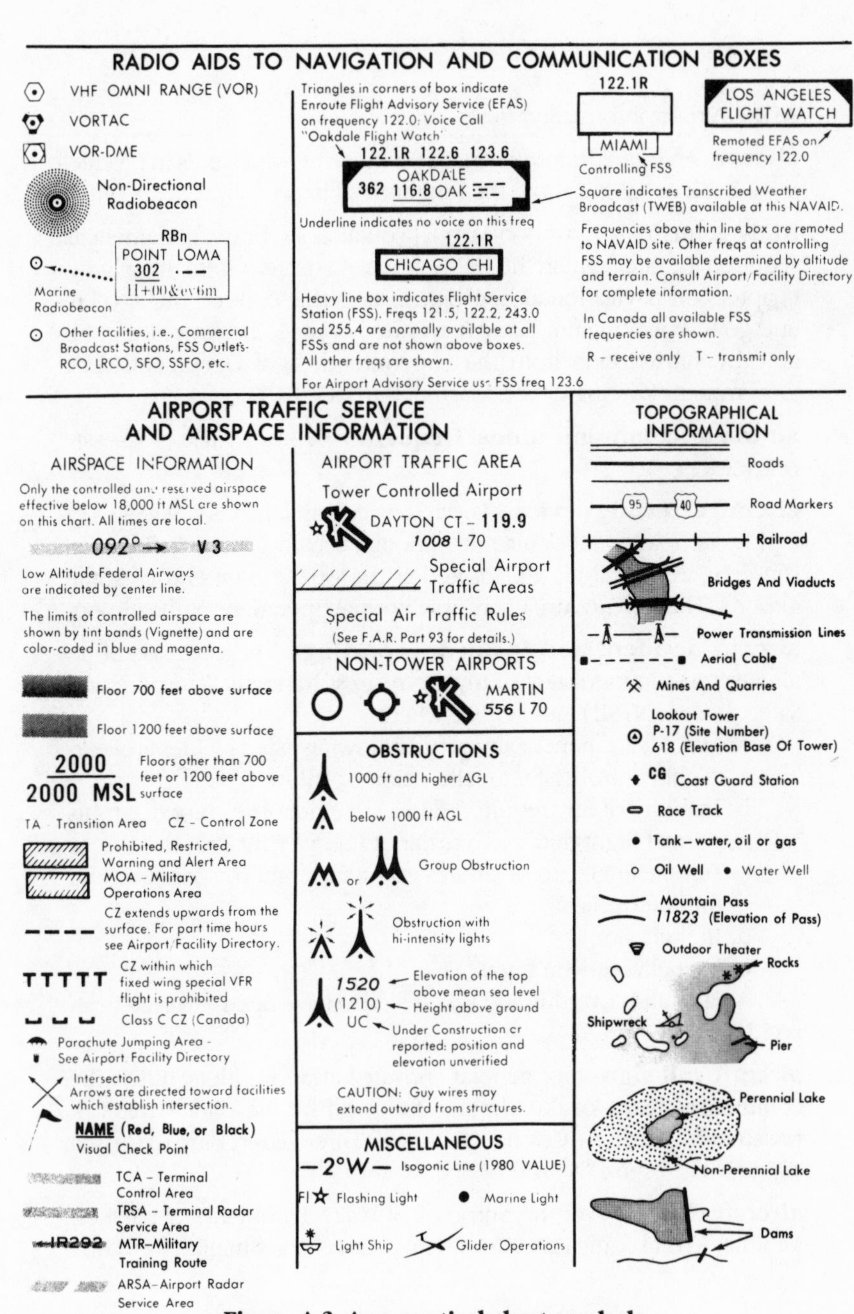

RADIO AIDS TO NAVIGATION AND COMMUNICATION BOXES
VHF OMNI RANGE (VOR)
VORTAC
VOR-DME
Non-Directional Radiobeacon
RBn
POINT LOMA
302
Marine Radiobeacon
Other facilities, i.e., Commercial Broadcast Stations, FSS Outlets-RCO, LRCO, SFO, SSFO, etc.
Triangles in corners of box indicate Enroute Flight Advisory Service (EFAS) on frequency 122.0: Voice Call "Oakdale Flight Watch"
122.1R 122.6 123.6
OAKDALE
362 116.8 OAK
Underline indicates no voice on this freq
122.1R
CHICAGO CHI
Heavy line box indicates Flight Service Station (FSS). Freqs 121.5, 122.2, 243.0 and 255.4 are normally available at all FSSs and are not shown above boxes. All other freqs are shown.
For Airport Advisory Service use FSS freq 123.6
122.1R
MIAMI
Controlling FSS
LOS ANGELES FLIGHT WATCH
Remoted EFAS on frequency 122.0
Square indicates Transcribed Weather Broadcast (TWEB) available at this NAVAID.
Frequencies above thin line box are remoted to NAVAID site. Other freqs at controlling FSS may be available determined by altitude and terrain. Consult Airport/Facility Directory for complete information.
In Canada all available FSS frequencies are shown.
R – receive only T – transmit only
AIRPORT TRAFFIC SERVICE AND AIRSPACE INFORMATION
AIRSPACE INFORMATION
Only the controlled and reserved airspace effective below 18,000 ft MSL are shown on this chart. All times are local.
092° V 3
Low Altitude Federal Airways are indicated by center line.
The limits of controlled airspace are shown by tint bands (Vignette) and are color-coded in blue and magenta.
Floor 700 feet above surface
Floor 1200 feet above surface
2000
2000 MSL
Floors other than 700 feet or 1200 feet above surface
TA - Transition Area CZ - Control Zone
Prohibited, Restricted, Warning and Alert Area
MOA - Military Operations Area
CZ extends upwards from the surface. For part time hours see Airport/Facility Directory.
TTTTT
CZ within which fixed wing special VFR flight is prohibited
Class C CZ (Canada)
Parachute Jumping Area - See Airport/Facility Directory
Intersection
Arrows are directed toward facilities which establish intersection.
NAME (Red, Blue, or Black)
Visual Check Point
TCA – Terminal Control Area
TRSA – Terminal Radar Service Area
IR292
MTR–Military Training Route
ARSA–Airport Radar Service Area
AIRPORT TRAFFIC AREA
Tower Controlled Airport
DAYTON CT – 119.9
1008 L 70
Special Airport Traffic Areas
Special Air Traffic Rules
(See F.A.R. Part 93 for details.)
NON-TOWER AIRPORTS
MARTIN
556 L 70
OBSTRUCTIONS
1000 ft and higher AGL
below 1000 ft AGL
or
Group Obstruction
Obstruction with hi-intensity lights
1520
(1210)
UC
Elevation of the top above mean sea level
Height above ground
Under Construction or reported: position and elevation unverified
CAUTION: Guy wires may extend outward from structures.
MISCELLANEOUS
–2°W–
Isogonic Line (1980 VALUE)
Fl
Flashing Light
Marine Light
Light Ship
Glider Operations
TOPOGRAPHICAL INFORMATION
Roads
95
40
Road Markers
Railroad
Bridges And Viaducts
Power Transmission Lines
Aerial Cable
Mines And Quarries
Lookout Tower
P-17 (Site Number)
618 (Elevation Base Of Tower)
CG
Coast Guard Station
Race Track
Tank – water, oil or gas
Oil Well
Water Well
Mountain Pass
11823 (Elevation of Pass)
Outdoor Theater
Rocks
Shipwreck
Pier
Perennial Lake
Non-Perennial Lake
Dams

Figure A-3. Aeronautical chart symbols.

AID—Airport information desk.

aileron—A hinged control surface on the wing which aids in producing a bank or rolling about the longitudinal axis.

AIM—Airman's Information Manual contains the basic fundamentals required for safe flight in the U.S. National Airspace System. It includes chapters on navigation aids, airspace, air traffic control, flight safety and good operating practices. It is issued every 112 days and is available by subscription from the Superintendent of Documents, U.S. Government Printing Office, Washington, DC 20402.

air-to-air communications frequency—122.75 MHz. (*See* FREQUENCIES.)

aircraft arresting device—Located on certain airports as a means of rapidly stopping military aircraft. Although usually located in the overrun area, a few cables are located at the operational end of some airports. They do not usually restrict normal operations at the airport.

aircraft accident and incident reporting—The operator of an aircraft must immediately notify the nearest National Transportation Safety Board (NTSB) Field Office when:

1. An aircraft accident or any of the following listed incidents occur:
   a. Flight control system malfunction or failure;
   b. Inability of any required flight crew member to perform his normal flight duties as a result of injury or illness;
   c. Turbine engine rotor failures excluding compressor blades and turbine buckets;
   d. In-flight fire;
   e. Aircraft collide in flight.
2. An aircraft is overdue and is believed to have been involved in an accident.

aircraft call sign—For general (private) aviation pilots, this is the commercial make of the airplane followed by the digits/letters of registration, such as Cessna One Two Three Four Bravo (Cessna 1234B). The prefix "N" is dropped.

aircraft classes—For the purposes of Wake Turbulence Separation Minima, ATC classifies aircraft as Heavy, Large, and Small as follows:

1. Heavy—Aircraft capable of take-off weights of 300,000 pounds or more whether or not they are operating at this weight during a particular phase of flight.
2. Large—Aircraft of more than 12,500 pounds, maximum certificated takeoff weight, up to 300,000 pounds.
3. Small—Aircraft of 12,500 pounds or less, maximum certificated takeoff weight. (*See* AIM.)

aircraft climbing/descending—Whenever cleared by air traffic control to change altitude, do so promptly on acknowledgment of the clearance. If the altitude change is 1,000 feet or less, descend or climb at the rate of 500 feet per minute; if the change is more than 1,000 feet, climb or descend as rapidly as practicable to 1,000 feet above or below the assigned altitude and then 500 feet per minute until the assigned altitude is reached.

aircraft conflict advisory—Immediately issued to an aircraft under his control by a controller if he is aware of an aircraft not under his control at an altitude believed to place the aircraft in unsafe proximity to each other. With the alert, he offers the pilot an alternative, if feasible.

aircraft equipment—(*See* VISUAL FLIGHT RULES, REQUIRED EQUIPMENT, DAY and VISUAL FLIGHT RULES, REQUIRED EQUIPMENT, NIGHT.)

aircraft operation lights—
1. FAA has initiated a voluntary pilot safety program, "Operation Lights On" to enhance the "see-and-be-seen" concept of averting collisions both in the air and on the ground, and to reduce bird strikes. All pilots are encouraged to turn on their anticollision lights any time the engine(s) are running day or night. All pilots are further encouraged to turn on their landing lights when operating within 10 miles of any airport (day and night), in conditions of reduced visibility and in areas where flocks of birds may be expected, i.e., coastal areas, lake areas, around refuse dumps, etc.
2. Although turning on aircraft lights does enhance the "see-and-be-seen" concept, pilots should not become complacent about keeping a sharp lookout for other aircraft. Not all aircraft are equipped with lights and some pilots may not have their lights turned on. The

aircraft operation lights ***continued***

aircraft manufacturers' recommendations for operation of landing lights and electrical systems should be observed. Pilots are reminded to extinguish aircraft strobe lights when on the ground, because of the flash intensity and irritating effect on ground personnel and other pilots.

aircraft performance information—(*See* owner's manual for each individual aircraft to be operated.)

aircraft radio receiving frequencies—*En route* communications from 108.0 through 117.95 MHz received on the VOR receiver or the localizer receiver; *air traffic control* communications from 118.0 through 121.4 MHz; *emergency* frequency, 121.5 MHz; ground control uses 121.6 to 121.9 MHz; aeronautical advisory stations (UNICOM) use 122.7, 122.8, and 123.0 MHz.

aircraft radio transmitting frequencies—Private aircraft to airport towers, 118.0 through 121.4 MHz (where the pilot can receive and transmit both); plus 122.5, 122.7, and 122.9 MHz where the pilot transmits on this frequency but receives on the published frequency for the tower; private or commercial aircraft to Federal Aviation Administration communications stations, 122.1 and 123.6.

aircraft use in the United States—There are approximately 210,000 general aviation aircraft of which 64,000 are used for business, 95,000 for personal reasons, 15,000 for instruction and the remainder for various work uses. There are 4,000 air carriers in the U.S. today.

air defense identification zone (ADIZ)—That area of land or water over which all aircraft must be readily identified, located, and controlled, in the interest of national security (*See* Figure A-4.)

air density—The weight of air by volume, which is affected by pressure, temperature, and humidity.

airfoil—The shape of the wings over which the air flows to provide maximum lift with minimum drag.

airframe icing—(*See* ICING, AIRFRAME.)

Airman's Information Manual (AIM)—A publication containing basic flight information and ATC procedures designed primarily as a

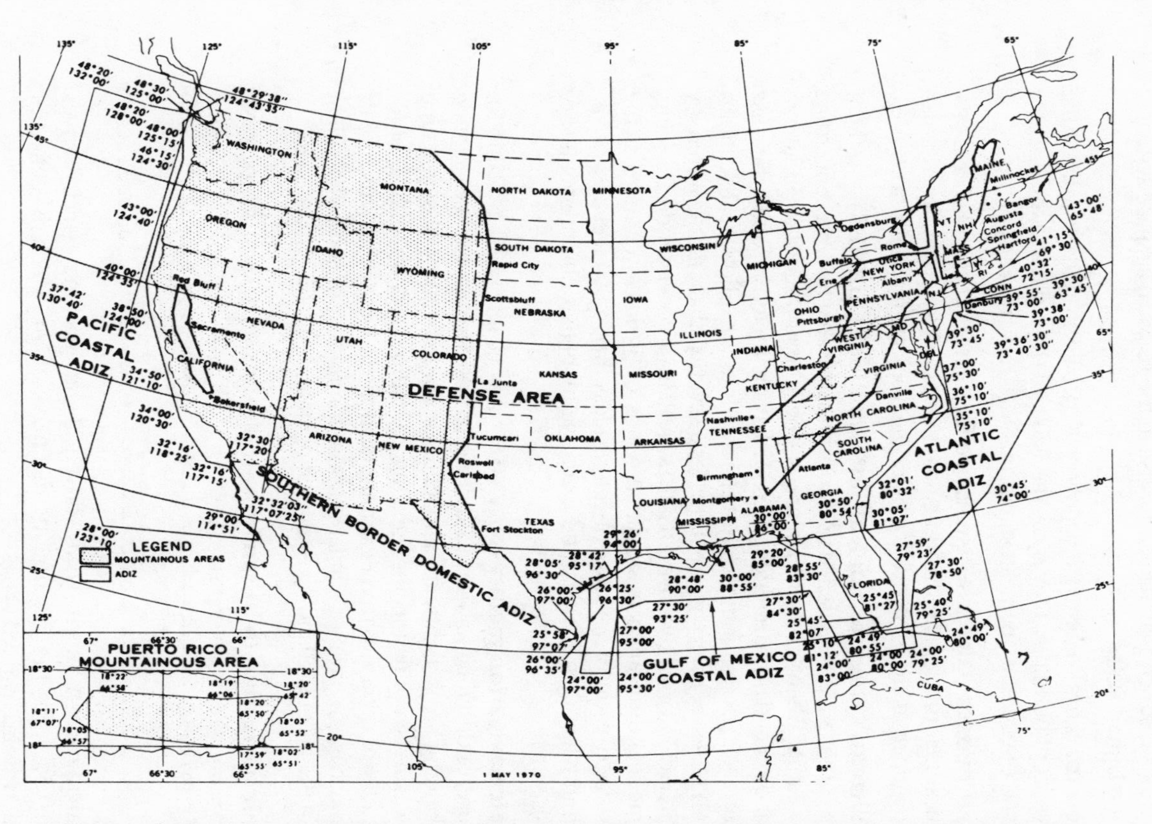

Figure A-4. Air defense identification zones, defense area, and designated mountainous areas.

Airman's Information Manual (AIM) ***continued***

pilot's instructional manual for use in the national airspace system of the United States.

air markers—Painted in yellow on dark backgrounds, usually on the roofs of buildings, they show the name of the town and the distance and direction to the nearest airport. Only a few remain.

air masses—Air masses are named for their sources and characteristics: maritime polar, maritime arctic (moist cold), continental polar, continental arctic (dry cold), maritime tropical (moist warm).

airman's meteorological information (AIRMET)—In-flight weather advisories issued only to amend the area forecast concerning weather phenomena which are of operational interest to all aircraft and potentially hazardous to aircraft having limited capability because of lack of equipment, instrumentation, or pilot qualifications. AIRMETs concern weather of less severity than that covered by SIGMETs or Convective SIGMETs. AIRMETs cover moderate icing, moderate turbulence, sustained winds of 30 knots or more at the surface, widespread areas of ceilings less than 1,000 feet and/or visibility less than 3 miles, and extensive mountain obscurement. (*See* SIGMET.)

air navigation facility (NAVAID)—Any facility used, available for use, or designed for use in aid of air navigation. Such facilities include landing areas, lights, any apparatus or equipment for disseminating weather information, for signaling, for radio direction finding, or for radio or other electronic communication, and any other structure or mechanism having a similar purpose for guiding or controlling flight in the air or the landing or takeoff of aircraft.

air navigation light symbols—(*See* AERONAUTICAL CHART SYMBOLS).

Rotating Light ---------------- ☆   Flashing Light ------------ Fl☆

Light Ship ----------------   Marine Light ------------ ●

airplane—A mechanically driven flying machine which derives its lift from the reaction of the mass of air which is deflected downward by fixed wings.

airplane owner's flight manual—For a particular make and model of aircraft, the owner's flight manual shows the proper procedure for loading and other important flight characteristics of the craft.

airport—A landing field for aircraft. There are approximately 16,000 airports in the United States, including heliports and seaplane bases. Commercial airlines serve approximately 700 airports. Most of the 11,000 privately owned and operated airports are not open to regular landings by the private aviation pilot. Check your chart.

airport advisory area—The area within 10 miles of an airport without a control tower or where the tower is not in operation and on which a flight service station is located. (*See* AIRPORT ADVISORY SERVICE.)

airport advisory service (AAS)—A service provided by Flight Service Stations at airports not served by a control tower. This service consists of providing information to arriving and departing aircraft concerning wind direction and speed, favored runway, altimeter setting, pertinent known traffic, pertinent known field conditions, airport taxi routes and traffic patterns, and authorized instrument approach procedures. This information is advisory in nature and does not constitute an ATC clearance. (*See* AIRPORT ADVISORY AREA.)

airport communications procedure—(See RADIO CONTACT PROCEDURE.)

airport elevation/field elevation—The highest point of an airport's usable runways measured in feet from mean sea level.

airport facility directory (AFD)—Published every 56 days by the National Ocean Survey. This publication lists airports, seaplane bases, and heliports open to the public; communications data, navigational facilities, and certain special notices, such as: parachute jumping, FSS/WX Service telephone numbers, Preferred Routes, and Aeronautical Chart Bulletins. Order information from National Ocean Survey, Distribution Division (C-44) Riverdale, MD 20737. Telephone (301) 436-6993 for subscription service.

airport information desk (AID)—A local airport unmanned facility designed for pilot self-service briefing, flight planning, and filing of flight plans.

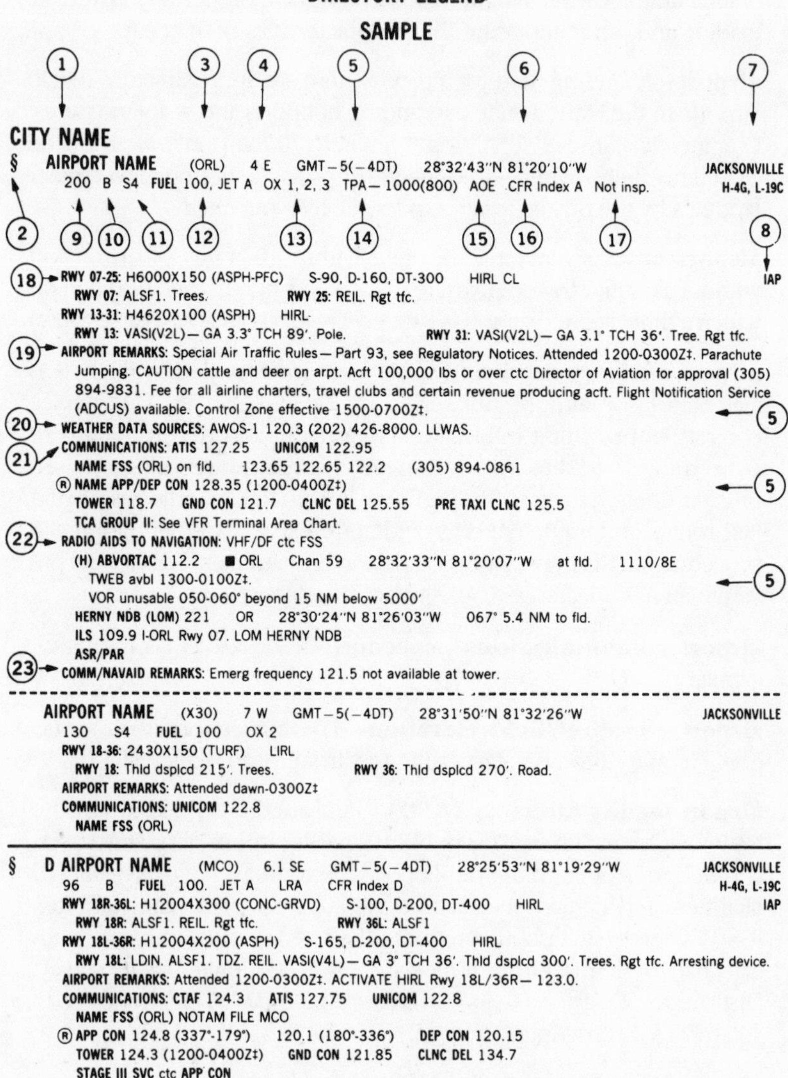
DIRECTORY LEGEND
SAMPLE
CITY NAME
§ AIRPORT NAME (ORL) 4 E GMT−5(−4DT) 28°32′43″N 81°20′10″W JACKSONVILLE
200 B S4 FUEL 100, JET A OX 1, 2, 3 TPA—1000(800) AOE CFR Index A Not insp. H-4G, L-19C
IAP
RWY 07-25: H6000X150 (ASPH-PFC) S-90, D-160, DT-300 HIRL CL
RWY 07: ALSF1. Trees. RWY 25: REIL. Rgt tfc.
RWY 13-31: H4620X100 (ASPH) HIRL
RWY 13: VASI(V2L)—GA 3.3° TCH 89′. Pole. RWY 31: VASI(V2L)—GA 3.1° TCH 36′. Tree. Rgt tfc.
AIRPORT REMARKS: Special Air Traffic Rules—Part 93, see Regulatory Notices. Attended 1200-0300Z‡. Parachute Jumping. CAUTION cattle and deer on arpt. Acft 100,000 lbs or over ctc Director of Aviation for approval (305) 894-9831. Fee for all airline charters, travel clubs and certain revenue producing acft. Flight Notification Service (ADCUS) available. Control Zone effective 1500-0700Z‡.
WEATHER DATA SOURCES: AWOS-1 120.3 (202) 426-8000. LLWAS.
COMMUNICATIONS: ATIS 127.25 UNICOM 122.95
NAME FSS (ORL) on fld. 123.65 122.65 122.2 (305) 894-0861
Ⓡ NAME APP/DEP CON 128.35 (1200-0400Z‡)
TOWER 118.7 GND CON 121.7 CLNC DEL 125.55 PRE TAXI CLNC 125.5
TCA GROUP II: See VFR Terminal Area Chart.
RADIO AIDS TO NAVIGATION: VHF/DF ctc FSS
(H) ABVORTAC 112.2 ■ ORL Chan 59 28°32′33″N 81°20′07″W at fld. 1110/8E
TWEB avbl 1300-0100Z‡.
VOR unusable 050-060° beyond 15 NM below 5000′
HERNY NDB (LOM) 221 OR 28°30′24″N 81°26′03″W 067° 5.4 NM to fld.
ILS 109.9 I-ORL Rwy 07. LOM HERNY NDB
ASR/PAR
COMM/NAVAID REMARKS: Emerg frequency 121.5 not available at tower.
1
2
3
4
5
6
7
8
9
10
11
12
13
14
15
16
17
18
19
20
21
22
23
AIRPORT NAME (X30) 7 W GMT−5(−4DT) 28°31′50″N 81°32′26″W JACKSONVILLE
130 S4 FUEL 100 OX 2
RWY 18-36: 2430X150 (TURF) LIRL
RWY 18: Thld dsplcd 215′. Trees. RWY 36: Thld dsplcd 270′. Road.
AIRPORT REMARKS: Attended dawn-0300Z‡
COMMUNICATIONS: UNICOM 122.8
NAME FSS (ORL)
§ D AIRPORT NAME (MCO) 6.1 SE GMT−5(−4DT) 28°25′53″N 81°19′29″W JACKSONVILLE
96 B FUEL 100. JET A LRA CFR Index D H-4G, L-19C
RWY 18R-36L: H12004X300 (CONC-GRVD) S-100, D-200, DT-400 HIRL IAP
RWY 18R: ALSF1. REIL. Rgt tfc. RWY 36L: ALSF1
RWY 18L-36R: H12004X200 (ASPH) S-165, D-200, DT-400 HIRL
RWY 18L: LDIN. ALSF1. TDZ. REIL. VASI(V4L)—GA 3° TCH 36′. Thld dsplcd 300′. Trees. Rgt tfc. Arresting device.
AIRPORT REMARKS: Attended 1200-0300Z‡. ACTIVATE HIRL Rwy 18L/36R—123.0.
COMMUNICATIONS: CTAF 124.3 ATIS 127.75 UNICOM 122.8
NAME FSS (ORL) NOTAM FILE MCO
Ⓡ APP CON 124.8 (337°-179°) 120.1 (180°-336°) DEP CON 120.15
TOWER 124.3 (1200-0400Z‡) GND CON 121.85 CLNC DEL 134.7
STAGE III SVC ctc APP CON

Figure A-5. Airport legend.

DIRECTORY LEGEND

RADIO AIDS TO NAVIGATION:
NAME (H) VORTAC 112.2 ORL Chan 59 28°32'33"N 81°20'07"W 173° 5.7 NM to fld. 1110/8E
MLS Chan 514 Rwy 36R

E AIRPORT NAME (See PLYMOUTH)

All Bearings and Radials are Magnetic unless otherwise specified.
All mileages are nautical unless otherwise noted.
All times are GMT except as noted.

This Directory is an alphabetical listing of data on record with the FAA on all airports that are open to the public, associated terminal control facilities, air route traffic control centers and radio aids to navigation within the conterminous United States, Puerto Rico and the Virgin Islands. Airports are listed alphabetically by associated city name and cross referenced by airport name. Facilities associated with an airport, but with a different name, are listed individually under their own name, as well as under the airport with which they are associated.
The listing of an airport in this directory merely indicates the airport operator's willingness to accommodate transient aircraft, and does not represent that the facility conforms with any Federal or local standards, or that it has been approved for use on the part of the general public.
The information on obstructions is taken from reports submitted to the FAA. It has not been verified in all cases. Pilots are cautioned that objects not indicated in this tabulation (or on charts) may exist which can create a hazard to flight operation.
Detailed specifics concerning services and facilities tabulated within this directory are contained in Airman's Information Manual, Basic Flight Information and ATC Procedures.
The legend items that follow explain in detail the contents of this Directory and are keyed to the circled numbers on the sample on the preceding page.

(1) CITY/AIRPORT NAME
Airports and facilities in this directory are listed alphabetically by associated city and state. Where the city name is different from the airport name the city name will appear on the line above the airport name. Airports with the same associated city name will be listed alphabetically by airport name and will be separated by a dashed rule line. All others will be separated by a solid rule line.

(2) NOTAM SERVICE
§—NOTAM "D" (Distant teletype dissemination) and NOTAM "L" (Local dissemination) service is provided for airport. Absence of annotation § indicates NOTAM "L" (Local dissemination) only is provided for airport. See AIM. Basic Flight Information and ATC Procedures for detailed description of NOTAM.

(3) LOCATION IDENTIFIER
A three or four character code assigned to airports. These identifiers are used by ATC in lieu of the airport name in flight plans, flight strips and other written records and computer operations.

(4) AIRPORT LOCATION
Airport location is expressed as distance and direction from the center of the associated city in nautical miles and cardinal points, i.e., 4 NE.

(5) TIME CONVERSION
Hours of operation of all facilities are expressed in Greenwich Mean Time (GMT) and shown as "Z" time. The directory indicates the number of hours to be subtracted from GMT to obtain local standard time and local daylight saving time GMT−5(−4DT). The symbol ‡ indicates that during periods of Daylight Saving Time effective hours will be one hour earlier than shown. In those areas where daylight saving time is not observed that (−4DT) and ‡ will not be shown. All states observe daylight savings time except Arizona and that portion of Indiana in the Eastern Time Zone and Puerto Rico and the Virgin Islands.

Figure A-5. Airport legend continued.

DIRECTORY LEGEND

(6) GEOGRAPHIC POSITION OF AIRPORT

(7) CHARTS

The Sectional Chart and Low and High Altitude Enroute Chart and panel on which the airport or facility is located.

(8) INSTRUMENT APPROACH PROCEDURES

IAP indicates an airport for which a prescribed (Public Use) FAA Instrument Approach Procedure has been published.

(9) ELEVATION

Elevation is given in feet above mean sea level and is the highest point on the landing surface. When elevation is sea level it will be indicated as (00). When elevation is below sea level a minus (–) sign will precede the figure.

(10) ROTATING LIGHT BEACON

B indicates rotating beacon is available. Rotating beacons operate dusk to dawn unless otherwise indicated in AIRPORT REMARKS.

(11) SERVICING

S1: Minor airframe repairs.
S2: Minor airframe and minor powerplant repairs.
S3: Major airframe and minor powerplant repairs.
S4: Major airframe and major powerplant repairs.

(12) FUEL

CODE	FUEL
80	Grade 80 gasoline (Red)
100	Grade 100 gasoline (Green)
100LL	Grade 100LL gasoline (low lead) (Blue)
115	Grade 115 gasoline
A	Jet A—Kerosene freeze point – 40° C.
A1	Jet A-1—Kerosene, freeze point – 50° C.
A1+	Jet A-1—Kerosene with icing inhibitor, freeze point – 50° C.
B	Jet B—Wide-cut turbine fuel, freeze point – 50° C.
B+	Jet B—Wide-cut turbine fuel with icing inhibitor, freeze point – 50° C.

(13) OXYGEN

OX 1 High Pressure
OX 2 Low Pressure
OX 3 High Pressure—Replacement Bottles
OX 4 Low Pressure—Replacement Bottles

(14) TRAFFIC PATTERN ALTITUDE

Traffic Pattern Altitude (TPA)—The first figure shown is TPA above mean sea level. The second figure in parentheses is TPA above airport elevation.

(15) AIRPORT OF ENTRY AND LANDING RIGHTS AIRPORTS

AOE—Airport of Entry—A customs Airport of Entry where permission from U.S. Customs is not required, however, at least one hour advance notice of arrival must be furnished.

LRA—Landing Rights Airport—Application for permission to land must be submitted in advance to U.S. Customs. At least one hour advance notice of arrival must be furnished.

NOTE: Advance notice of arrival at both an AOE and LRA airport may be included in the flight plan when filed in Canada or Mexico, where Flight Notification Service (ADCUS) is available the airport remark will indicate this service. This notice will also be treated as an application for permission to land in the case of an LRA. Although advance notice of arrival may be relayed to Customs through Mexico, Canadian, and U.S. Communications facilities by flight plan, the aircraft operator is solely responsible for insuring that Customs receives the notification. (See Customs, Immigration and Naturalization, Public Health and Agriculture Department requirements in the International Flight Information Manual for further details.)

Figure A-5. Airport legend continued.

DIRECTORY LEGEND

(16) CERTIFICATED AIRPORT (FAR 139)

Airports serving Civil Aeronautics Board certified carriers and certified under FAR, Part 139, are indicated by the CFR index; i.e., CFR Index A, which relates to the availability of crash, fire, rescue equipment.

FAR–PART 139 CERTIFICATED AIRPORTS

INDICES AND FIRE FIGHTING AND RESCUE EQUIPMENT REQUIREMENTS

Airport Index	*Required No. Vehicles*	*Aircraft Length*	*Scheduled Departures*	*Agent + Water for Foam*
A	1	≤90′	≥1	500#DC or 450#DC + 50 gal H_2O
AA	1	>90′, ≤126′	<5	300#DC + 500 gal H_2O
B	2	>90′, ≤126′	≥5	Index A + 1500 gal H_2O
		>126′, ≤160′	<5	
C	3	>126′, ≤160′	≥5	Index A + 3000 gal H_2O
		>160′, ≤200′	<5	
D	3	>160′, ≤200′	≥5	Index A + 4000 gal H_2O
		>200′	<5	
E	3	>200′	≥5	Index A + 6000 gal H_2O

> Greater Than; < Less Than; ≥ Equal or Greater Than; ≤ Equal or Less Than; H_2O–Water; DC–Dry Chemical.

NOTE: If AFFF (Aqueous Film Forming Foam) is used in lieu of Protein Foam, the water quantities listed for Indices AA thru E can be reduced 33 ⅓%. See FAR Part 139.49 for full details. The listing of CFR index does not necessarily assure coverage for non-air carrier operations or at other than prescribed times for air carrier. CFR index Ltd.—indicates CFR coverage may or may not be available, for information contact airport manager prior to flight.

(17) FAA INSPECTION

All airports not inspected by FAA will be identified by the note: Not insp. This indicates that the airport information has been provided by the owner or operator of the field.

(18) RUNWAY DATA

Runway information is shown on two lines. That information common to the entire runway is shown on the first line while information concerning the runway ends are shown on the second or following line. Lengthy information will be placed in the Airport Remarks.

Runway direction, surface, length, width, weight bearing capacity, lighting, gradient (when gradient exceeds 0.3 percent) and appropriate remarks are shown for each runway. Direction, length, width, lighting and remarks are shown for sealanes. The full dimensions of helipads are shown, i.e., 50X150.

RUNWAY SURFACE AND LENGTH

Runway lengths prefixed by the letter "H" indicate that the runways are hard surfaced (concrete, asphalt). If the runway length is not prefixed, the surface is sod, clay, etc. The runway surface composition is indicated in parentheses after runway length as follows:

(AFSC)—Aggregate friction seal coat
(ASPH)—Asphalt
(CONC)—Concrete
(DIRT)—Dirt
(GRVD)—Grooved
(GRVL)—Gravel, or cinders
(PFC)—Porous friction courses
(RFSC)—Rubberized friction seal coat
(TURF)—Turf
(TRTD)—Treated
(WC)—Wire combed

Figure A-5. Airport legend continued.

DIRECTORY LEGEND

RUNWAY WEIGHT BEARING CAPACITY

Runway strength data shown in this publication is derived from available information and is a realistic estimate of capability at an average level of activity. It is not intended as a maximum allowable weight or as an operating limitation. Many airport pavements are capable of supporting limited operations with gross weights of 25-50% in excess of the published figures. Permissible operating weights, insofar as runway strengths are concerned, are a matter of agreement between the owner and user. When desiring to operate into any airport at weights in excess of those published in the publication, users should contact the airport management for permission. Add 000 to figure following S, D, DT, DDT and MAX for gross weight capacity:

S—Runway weight bearing capacity for aircraft with single-wheel type landing gear, (DC-3), etc.

D—Runway weight bearing capacity for aircraft with dual-wheel type landing gear, (DC-6), etc.

DT—Runway weight bearing capacity for aircraft with dual-tandem type landing gear, (707), etc.

DDT—Runway weight bearing capacity for aircraft with double dual-tandem type landing gear, (747), etc.

Quadricycle and dual-tandem are considered virtually equal for runway weight bearing consideration, as are single-tandem and dual-wheel.

Omission of weight bearing capacity indicates information unknown.

RUNWAY LIGHTING

Lights are in operation sunset to sunrise. Lighting available by prior arrangement only or operating part of the night only and/or pilot controlled and with specific operating hours are indicated under airport remarks. Since obstructions are usually lighted, obstruction lighting is not included in this code. Unlighted obstructions on or surrounding an airport will be noted in airport remarks.

Temporary, emergency or limited runway edge lighting such as flares, smudge pots, lanterns or portable runway lights will also be shown in airport remarks.

Types of lighting are shown with the runway or runway end they serve.

LIRL—Low Intensity Runway Lights
MIRL—Medium Intensity Runway Lights
HIRL—High Intensity Runway Lights
REIL—Runway End Identifier Lights
CL—Centerline Lights
TDZ—Touchdown Zone Lights
ODALS—Omni Directional Approach Lighting System.
AF OVRN—Air Force Overrun 1000' Standard Approach Lighting System.
LDIN—Lead-In Lighting System.
MALS—Medium Intensity Approach Lighting System.
MALSF—Medium Intensity Approach Lighting System with Sequenced Flashing Lights.
MALSR—Medium Intensity Approach Lighting System with Runway Alignment Indicator Lights.
SALS—Short Approach Lighting System.
SALSF—Short Approach Lighting System with Sequenced Flashing Lights.
SSALS—Simplified Short Approach Lighting System.
SSALF—Simplified Short Approach Lighting System with Sequenced Flashing Lights.
SSALR—Simplified Short Approach Lighting System with Runway Alignment Indicator Lights.
ALSAF—High Intensity Approach Lighting System with Sequenced Flashing Lights
ALSFI—High Intensity Approach Lighting System with Sequenced Flashing Lights, Category I, Configuration.
ALSF2—High Intensity Approach Lighting System with Sequenced Flashing Lights, Category II, Configuration.
VASI—Visual Approach Slope Indicator System.

VISUAL APPROACH SLOPE INDICATOR SYSTEMS

VASI—Visual Approach Slope Indicator
SAVASI—Simplified Abbreviated Visual Approach Slope Indicator

S2L 2-box SAVASI on left side of runway
S2R 2-box SAVASI on right side of runway
V2R 2-box VASI on right side of runway
V2L 2-box VASI on left side of runway
V4R 4-box VASI on right side of runway
V4L 4-box VASI on left side of runway
V6R 6-box VASI on right side of runway
V6L 6-box VASI on left side of runway
V12 12-box VASI on both sides of runway
V16 16-box VASI on both sides of runway
*NSTD Nonstandard VASI, VAPI, or any other system not listed above

Figure A-5. Airport legend continued.

DIRECTORY LEGEND

VASI approach slope angle and threshold crossing height will be shown when available; i.e., GA 3.5° TCH 37.0′.

PILOT CONTROL OF AIRPORT LIGHTING

Key Mike	Function
7 times within 5 seconds	Highest intensity available
5 times within 5 seconds	Medium or lower intensity (Lower REIL or REIL-Off)
3 times within 5 seconds	Lowest intensity available (Lower REIL or REIL-Off)

Available systems will be indicated in the Airport Remarks, as follows:

ACTIVATE MALSR Rwy 7, HIRL Rwy 7/25-122.8.
or
ACTIVATE MIRL Rwy 18/36-122.8.
or
ACTIVATE VASI and REIL, Rwy 7-122.8.

Where the airport is not served by an instrument approach procedure and/or has an independent type system of different specification installed by the airport sponsor, descriptions of the type lights, method of control, and operating frequency will be explained in clear text. See AIM, "Basic Flight Information and ATC Procedures," for detailed description of pilot control of airport lighting.

RUNWAY GRADIENT

Runway gradient will be shown only when it is 0.3 percent or more. When available the direction of slope upward will be indicated, i.e., 0.5% up NW.

RUNWAY END DATA

Lighting systems such as VASI, MALSR, REIL; obstructions; displaced thresholds will be shown on the specific runway end. "Rgt tfc"—Right traffic indicates right turns should be made on landing and takeoff for specified runway end.

(19) AIRPORT REMARKS

Landing Fee indicates landing charges for private or non-revenue producing aircraft, in addition, fees may be charged for planes that remain over a couple of hours and buy no services, or at major airline terminals for all aircraft.
Remarks—Data is confined to operational items affecting the status and usability of the airport.

(20) WEATHER DATA SOURCES

AWOS—Automated Weather Observation System

AWOS-1—reports altimeter setting, wind data and usually temperature, dewpoint and density altitude.
AWOS-2—reports the same as AWOS-1 plus visibility.
AWOS-3—reports the same as AWOS-1 plus visibility and cloud/ceiling data.
See AIM, Basic Flight Information and ATC Procedures for detailed description of AWOS.

SAWRS—identifies airports that have a Supplemental Aviation Weather Reporting Station available to pilots for current weather information.
LLWAS—indicates a Low Level Wind Shear Alert System consisting of a center field and several field perimeter anemometers.

(21) COMMUNICATIONS

Communications will be listed in sequence in the order shown below:
Common Traffic Advisory Frequency (CTAF), Automatic Terminal Information Service (ATIS) and Aeronautical Advisory Stations (UNICOM) along with their frequency is shown, where available, on the line following the heading "COMMUNICATIONS." When the CTAF and UNICOM is the same frequency, the frequency will be shown as CTAF/UNICOM freq.
Flight Service Station (FSS) information. The associated FSS will be shown followed by the identifier and information concerning availablity of telephone service, e.g. Direct Line (DL), Local Call (LC), etc. Where the airport NOTAM File identifier is different than the associated FSS it will be shown as "NOTAM FILE IAD." Where the FSS is located on the field it will be indicated as "on arpt"

Figure A-5. Airport legend continued.

following the identifier. Frequencies available will follow. The FSS telephone number will follow along with any significant operational information. FSS's whose name is not the same as the airport on which located will also be listed in the normal alphabetical name listing for the state in which located. Remote Communications Outlet (RCO) providing service to the airport followed by the frequency and name of the Controlling FSS.

FSS's provide information on airport conditions, radio aids and other facilities, and process flight plans. Airport Advisory Service is provided on the CTAF by FSS's located at non-tower airports or airports where the tower is not in operation.

(See AIM, Par. 157/158 Traffic Advisory Practices at airports where a tower is not in operation or AC 90 - 42C.)

Aviation weather briefing service is provided by FSS specialists. Flight and weather briefing services are also available by calling the telephone numbers listed.

Remote Communications Outlet (RCO)—An unmanned air/ground communications facility, remotely controlled and providing UHF or VHF communications capability to extend the service range of an FSS.

Civil Communications Frequencies—Civil communications frequencies used in the FSS air/ground system are now operated simplex on 122.0, 122.2, 122.3, 122.4, 122.6, 123.6; emergency 121.5; plus receive-only on 122.05, 122.1, 122.15, and 123.6.

a. 122.0 is assigned as the Enroute Flight Advisory Service channel at selected FSS's.
b. 122.2 is assigned to all FSS's as a common enroute simplex service.
c. 123.6 is assigned as the airport advisory channel at non-tower FSS locations, however, it is still in commission at some FSS's collocated with towers to provide part time Airport Advisory Service.
d. 122.1 is the primary receive-only frequency at VOR's. 122.05, 122.15 and 123.6 are assigned at selected VOR's meeting certain criteria.
e. Some FSS's are assigned 50 kHz channels for simplex operation in the 122-123 MHz band (e.g. 122.35). Pilots using the FSS A/G system should refer to this directory or appropriate charts to determine frequencies available at the FSS or remoted facility through which they wish to communicate.

Part time FSS hours of operation are shown in remarks under facility name.

Emergency frequency 121.5 is available at all Flight Service Stations, Towers, Approach Control and RADAR facilities, unless indicated as not available.

Frequencies published followed by the letter "T" or "R", indicate that the facility will only transmit or receive respectively on that frequency. All radio aids to navigation frequencies are transmit only.

TERMINAL SERVICES

CTAF—A program designed to get all vehicles and aircraft at uncontrolled airports on a common frequency.
ATIS—A continuous broadcast of recorded non-control information in selected areas of high activity.
UNICOM—A non-government air/ground radic communications facility utilized to provide general airport advisory service.
APP CON—Approach Control. The symbol Ⓡ indicates radar approach control.
TOWER—Control tower
GND CON—Ground Control
DEP CON—Departure Control. The symbol Ⓡ indicates radar departure control.
CLNC DEL—Clearance Delivery.
PRE TAXI CLNC—Pre taxi clearance
VFR ADVSY SVC—VFR Advisory Service. Service provided by Non-Radar Approach Control.
Advisory Service for VFR aircraft (upon a workload basis) ctc APP CON.
STAGE II SVC—Radar Advisory and Sequencing Service for VFR aircraft
STAGE III SVC—Radar Sequencing and Separation Service for participating VFR Aircraft within a Terminal Radar Service Area (TRSA)
ARSA—Airport Radar Service Area
TCA—Radar Sequencing and Separation Service for all aircraft in a Terminal Control Area (TCA)
TOWER, APP CON and DEP CON RADIO CALL will be the same as the airport name unless indicated otherwise.

(22) RADIO AIDS TO NAVIGATION

The Airport Facility Directory lists by facility name all Radio Aids to Navigation, except Military TACANS, that appear on National Ocean Service Visual or IFR Aeronautical Charts and those upon which the FAA has approved an Instrument Approach Procedure. All VOR, VORTAC ILS and MLS equipment in the National Airspace System has an automatic monitoring and shutdown feature in the event of malfunction. Unmonitored, as used in this publication for any navigational aid, means that FSS or tower personnel cannot observe the malfunction or shutdown signal.

Figure A-5. Airport legend continued.

DIRECTORY LEGEND

NAVAID information is tabulated as indicated in the following sample:

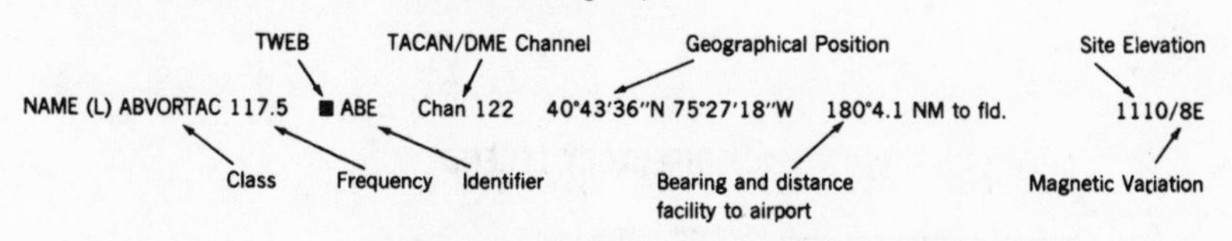

VOR unusable 020°-060° beyond 26 NM below 3500°

Restriction within the normal altitude/range of the navigational aid (See primary alphabetical listing for restrictions on VORTAC and VOR/DME).

ASR/PAR—Indicates that Surveillance (ASR) or Precision (PAR) radar instrument approach minimums are published in U.S. Government Instrument Approach Procedures.

RADIO CLASS DESIGNATIONS

Identification of VOR/VORTAC/TACAN Stations by Class (Operational Limitations):

Normal Usable Altitudes and Radius Distances

Class	Altitudes	Distance (miles)
(T)	12,000' and below	25
(L)	Below 18,000'	40
(H)	Below 18,000'	40
(H)	Within the Conterminous 48 States only, between 14,500' and 17,999'	100
(H)	18,000' FL 450	130
(H)	Above FL 450	100

(H) = High (L) = Low (T) = Terminal

NOTE: An (H) facility is capable of providing (L) and (T) service volume and an (L) facility additionally provides (T) service volume.

The term VOR is, operationally, a general term covering the VHF omnidirectional bearing type of facility without regard to the fact that the power, the frequency protected service volume, the equipment configuration, and operational requirements may vary between facilities at different locations.

AB ———— Automatic Weather Broadcast (also shown with ■ following frequency.)
DF ———— Direction Finding Service.
DME ———— UHF standard (TACAN compatible) distance measuring equipment.
H ———— Non-directional radio beacon (homing), power 50 watts to less than 2,000 watts (50 NM at all altitudes).
HH ———— Non-directional radio beacon (homing), power 2,000 watts or more (75 NM at all altitudes).
H-SAB ———— Non-directional radio beacons providing automatic transcribed weather service.
ILS ———— Instrument Landing System (voice, where available, on localizer channel).
ISMLS ———— Interim Standard Microwave Landing System.
LDA ———— Localizer Directional Aid.
LMM ———— Compass locator station when installed at middle marker site (15 NM at all altitudes).
LOM ———— Compass locator station when installed at outer marker site (15 NM at all altitudes).
MH ———— Non-directional radio beacon (homing) power less than 50 watts (25 NM at all altitudes).

Figure A-5. Airport legend continued.

DIRECTORY LEGEND

MLS	Microwave Landing System
S	Simultaneous range homing signal and/or voice.
SABH	Non-directional radio beacon not authorized for IFR or ATC. Provides automatic weather broadcasts.
SDF	Simplified Direction Facility.
TACAN	UHF navigational facility-omnidirectional course and distance information.
VOR	VHF navigational facility-omnidirectional course only.
VOR/DME	Collocated VOR navigational facility and UHF standard distance measuring equipment.
VORTAC	Collocated VOR and TACAN navigational facilities.
W	Without voice on radio facility frequency.
Z	VHF station location marker at a LF radio facility.

FREQUENCY PAIRING PLAN

The following is a list of paired VOR/ILS VHF frequencies with TACAN Channels:

Frequency MHz	Channel	Frequency MHz	Channel	Frequency MHz	Channel	Frequency MHz	Channel
108.0	17	110.5	42	113.0	77	115.5	102
108.1	18	110.6	43	113.1	78	115.6	103
108.2	19	110.7	44	113.2	79	115.7	104
108.3	20	110.8	45	113.3	80	115.8	105
108.4	21	110.9	46	113.4	81	115.9	106
108.5	22	111.0	47	113.5	82	116.0	107
108.6	23	111.1	48	113.6	83	116.1	108
108.7	24	111.2	49	113.7	84	116.2	109
108.8	25	111.3	50	113.8	85	116.3	110
108.9	26	111.4	51	113.9	86	116.4	111
109.0	27	111.5	52	114.0	87	116.5	112
109.1	28	111.6	53	114.1	88	116.6	113
109.2	29	111.7	54	114.2	89	116.7	114
109.3	30	111.8	55	114.3	90	116.8	115
109.4	31	111.9	56	114.4	91	116.9	116
109.5	32	112.0	57	114.5	92	117.0	117
109.6	33	112.1	58	114.6	93	117.1	118
109.7	34	112.2	59	114.7	94	117.2	119
109.8	35	112.3	70	114.8	95	117.3	120
109.9	36	112.4	71	114.9	96	117.4	121
110.0	37	112.5	72	115.0	97	117.5	122
110.1	38	112.6	73	115.1	98	117.6	123
110.2	39	112.7	74	115.2	99	117.7	124
110.3	40	112.8	75	115.3	100	117.8	125
110.4	41	112.9	76	115.4	101	117.9	126

Figure A-5. Airport legend continued.

MLS CHANNELING

The following is a list of MLS channels and frequencies. MLS has 200 channels available.

Channel Number	Frequency (MHz)	Channel Number	Frequency (MHz)	Channel Number	Frequency (MHz)	Channel Number	Frequency (MHz)	Channel Number	Frequency (MHz)
500	5031.0	540	5043.0	580	5055.0	620	5067.0	660	5079.0
501	5031.3	.	.	.	.	.	.	.	.
502	5031.6	.	.	.	.	.	.	.	.
503	5031.9	.	.	.	.	.	.	.	.
504	5032.2	.	.	.	.	.	.	.	.
505	5032.5	545	5044.5	585	5056.5	625	5068.5	665	5080.5
.	.	.	.	.	.	.	.	.	.
.	.	.	.	.	.	.	.	.	.
.	.	.	.	.	.	.	.	.	.
.	.	.	.	.	.	.	.	.	.
510	5034.0	550	5046.0	590	5058.0	630	5070.0	670	5082.0
.	.	.	.	.	.	.	.	.	.
.	.	.	.	.	.	.	.	.	.
.	.	.	.	.	.	.	.	.	.
.	.	.	.	.	.	.	.	.	.
515	5035.5	555	5047.5	595	5059.5	635	5071.5	675	5083.5
.	.	.	.	.	.	.	.	.	.
.	.	.	.	.	.	.	.	.	.
.	.	.	.	.	.	.	.	.	.
.	.	.	.	.	.	.	.	.	.
520	5037.0	560	5049.0	600	5061.0	640	5073.0	680	5085.0
.	.	.	.	.	.	.	.	.	.
.	.	.	.	.	.	.	.	.	.
.	.	.	.	.	.	.	.	.	.
.	.	.	.	.	.	.	.	.	.
525	5038.5	565	5050.5	605	5062.5	645	5074.5	685	5086.5
.	.	.	.	.	.	.	.	.	.
.	.	.	.	.	.	.	.	.	.
.	.	.	.	.	.	.	.	.	.
.	.	.	.	.	.	.	.	.	.
530	5040.0	570	5052.0	610	5064.0	650	5076.0	690	5088.0
.	.	.	.	.	.	.	.	.	.
.	.	.	.	.	.	.	.	.	.
.	.	.	.	.	.	.	.	.	.
.	.	.	.	.	.	.	.	.	.
535	5041.5	575	5053.5	615	5065.5	655	5077.5	695	5089.5
.	.	.	.	.	.	.	.	696	5089.8
.	.	.	.	.	.	.	.	697	5090.1
.	.	.	.	.	.	.	.	698	5090.4
.	.	.	.	.	.	.	.	699	5090.7

(23) COMM/NAVAID REMARKS:
Pertinent remarks concerning communications and NAVAIDS.

Figure A-5. Airport legend continued.

airport lighting—(*See* APPROACH LIGHT SYSTEM, VISUAL APPROACH SLOPE INDICATOR [VASI]).

airport lighting, pilot control of—Radio control of lighting is available at selected airports to provide airborne or ground control of lights by keying the aircraft's microphone. Control of lighting systems is available either full-time at those locations without specified hours for lighting or with no control tower or FSS, or part-time (during unmanned periods) at those locations with a part-time tower or FSS or specified hours for lighting. All lighting systems which are to be radio controlled at an airport, whether on a single runway or multiple runways, operate on the same radio frequency.

Suggested use is to always initially key the mike 7 times; this assures that all controlled lights are turned on to the maximum available intensity. If desired, adjustment can then be made, where the capability is provided, to a lower intensity (or the REIL turned off) by keying 5 and/or 3 times. Even though the runway lights are on when arriving over an airport of intended landing, always key mike as directed in order to assure a full 15 minutes lighting duration. Approved lighting systems may be activated by keying the mike (within 5 seconds) as indicated below:

AIRPORT LIGHTING AIDS

RADIO CONTROL SYSTEM

Key Mike	*Function*
7 times within 5 seconds	Highest intensity available
5 times within 5 seconds	Medium or lower intensity (Lower REIL or REIL-Off)
3 times within 5 seconds	Lowest intensity available (Lower REIL or REIL-Off)

For all public use airports with FAA standard systems the *Airport/Facility Directory* contains the types of lighting, runway, and the frequency that is used to activate the system. Airports with IAPs include data on the approach chart indentifying the light system, the runway on which they are installed, and the frequency that is used to activate the system.

Example: ACTIVATE MIRL Rwy 18/36-122.8.

Example: ACTIVATE MALSR, VASI and REIL Rwy 7, HIRL Rwy 7-25, and REIL Rwy 25-122.8.

Where the airport is not served by an IAP, it may have either the standard FAA approved control system or an independent type system of different specification installed by the airport sponsor. The *Airport/Facility Directory* contains descriptions of pilot controlled lighting systems for each airport having other than FAA approved systems, and explains the type lights, method of control, and operating frequency in clear text.

airport marking—(*See* RUNWAY MARKING.)

airport, operation in vicinity of—No person may operate an arriving aircraft below 10,000 feet MSL within an airport traffic area at an indicated airspeed of more than 200 knots (230 m.p.h.).

airport radar service area—An airport area in which all pilots are required to establish radio contact with air traffic controllers and abide by their instructions.

airport, rotating beacons—

1. The airport beacon has a vertical light distribution such as to make it most effective at angles of one up to 10 degrees above the horizontal from its site; however, it can be seen well above and below this peak spread. The beacon may be an omnidirectional capacitor-discharge device, or it may rotate at a constant speed which produces the visual effect of flashes at regular intervals. Flashes may be one or two colors alternately. The total number of flashes are:
   a. 12 to 30 per minute for beacons marking airports, landmarks, and points on Federal airways.
   b. 30 to 60 per minute for beacons marking heliports.
2. The colors and color combinations of beacons are:
   a. White and Green—Lighted land airport
   b. *Green alone—Lighted land airport
   c. White and Yellow—Lighted water airport
   d. *Yellow alone—Lighted water airport
   e. White and Red—Landmark or navigational point (rare installation)
   f. White alone—Unlighted land airport (rare installation)
   g. Green, Yellow, and White—Lighted heliport

*Green alone or yellow alone is used only in connection with a not

airport, rotating beacons ***continued***

far distant white-and-green or white-and-yellow beacon display, respectively.

3. Military airport beacons flash alternately white and green, but are differentiated from civil beacons by dual-peaked (two quick) white flashes between the green flashes.

4. In control zones, operation of the airport beacon during the hours of daylight often indicates that the ground visibility is less than 3 miles and/or the ceiling is less than 1,000 feet. ATC clearance in accordance with FAR 91 would be required for landing, takeoff, and flight in the traffic pattern. Pilots should not rely solely on the operation of the airport beacon to indicate weather conditions, IFR versus VFR. At locations with control towers and if controls are provided, ATC personnel turn the beacon on. However, at many airports throughout the country, the airport beacon is turned on by a photoelectric cell or time clocks and ATC personnel have no control as to when it shall be turned on. Also, there is no regulatory requirement for daylight operation and pilots are reminded that it remains their responsibility for complying with proper preflight planning in accordance with FAR—Part 91.

There are also rotating beacons and high-intensity lights on aircraft for better visibility of their movements at night.

airport servicing symbols—

S1: Minor air frame repairs
S2: Minor airframe and minor power-plant repairs
S3: Major airframe and minor power-plant repairs
S4: Major airframe and major power-plant repairs

airport surface detection equipment (ASDE)—Radar equipment specifically designed to detect all principal features on the surface of an airport, including vehicular traffic, and to present the entire picture on a radar indicator console in the control tower.

airport surveillance radar (ASR)—Approach control radar used to detect and display an aircraft's position in the terminal area. ASR provides range and azimuth information but does not provide elevation data. Coverage of the ASR can extend up to 60 miles.

airport symbols—(*See* Figure A-1.)

airport traffic area—Unless otherwise specifically designated (Federal Aviation Regulations, Part 93), this refers to the airspace within a horizontal radius of 5 statute miles from the geographical center of any airport at which a control tower is operating, extending from the surface up to, but not including, an altitude of 3,000 feet above the elevation of the airport. [*See* Figure A-6(a) and (b).]

airport traffic control service—Air traffic control service provided by an airport traffic control tower for aircraft operating on the movement area and in the vicinity of an airport.

airport traffic control tower (TOWER)—A facility providing airport traffic control service.

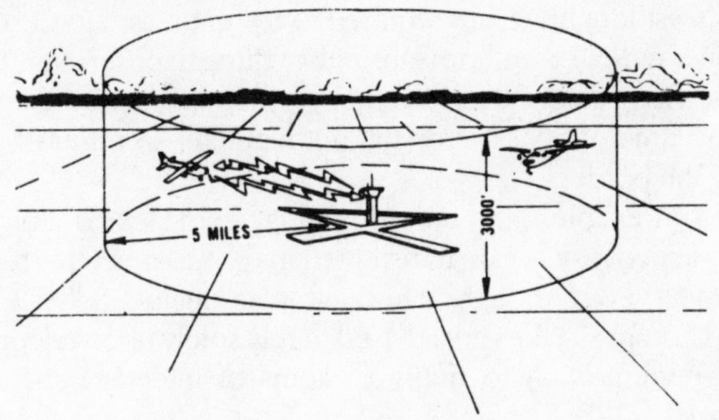

Figure A-6(a). Airport traffic area.

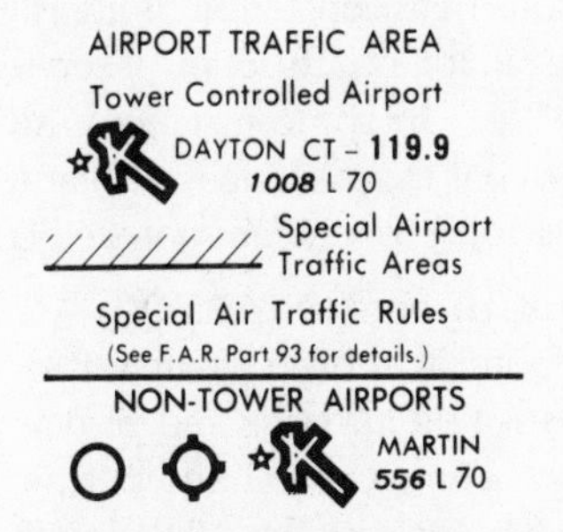

Figure A-6(b). Airport traffic area symbols. (The circles and information for airports such as Martin appear in magenta [purplish red] on aeronautical charts.)

airport utility radio frequency—The frequencies of 121.6 to 121.925 MHz are assigned for ground control use.

airports, nontower—(*See* NONTOWER AIRPORTS.)

airports of entry—Specific airports where a Customs office is nearby and no special permission is needed to land on arriving from outside the United States. However, one-hour advance notice must be furnished to United States Customs. (This can be included in your flight plan filed to Canada or Mexico, for example.)

All pilots of private aircraft arriving from a foreign place in the western hemisphere south of 33 degrees north latitude which cross into the U.S. over a point on the U.S. border between 95 and 120 degrees west longitude, are required to communicate to Customs by telephone, radio, or other means either directly or through the FAA Flight Service Station, their intention of landing and the intended point and time of border crossing not less than 15 minutes prior to crossing the border.

Due to unreliable communications relay from Mexico, a flight plan filed in Mexico may not be transmitted in time to meet this reporting requirement even though the flight plan included "ADCUS" or "Advise Customs." Pilots are advised to rely solely upon telephone or radio communication to notify Customs of intended arrival on a timely basis.

air route surveillance radar (ARSR)—Air route traffic control center (ARTCC) radar used primarily to detect and display an aircraft's position while en route between terminal areas. The ARSR enables controllers to provide radar air traffic control service when aircraft are within the ARSR coverage. In some instances, ARSR may enable an ARTCC to provide terminal radar services similar to but usually more limited than those provided by a radar approach control.

air route traffic control center (ARTCC)—A facility established to provide air traffic control service to aircraft operating on an IFR flight plan within controlled airspace and principally during the en route phase of flight. When capabilities permit, certain advisory/assistance services may be provided to VFR aircraft.

Air Safety Foundation—A nonprofit organization for aviation safety and research. It is supported solely by Aircraft Owners and Pilots

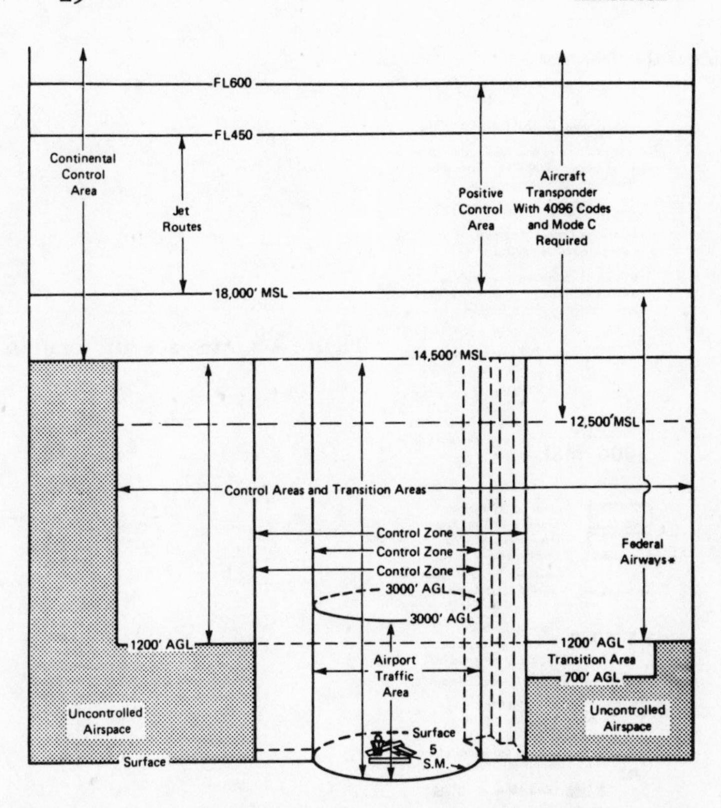

Figure A-7. Vertical extent of airspace segments. (NOTE: The above depicts the normal vertical limits of the various airspace segments.)

Association member donations, endowments, and so on.

airspace—(*See* Figures A-7 and A-8.)

airspeed—The speed of an airplane in relation to the air through which it is passing. The four types are detailed below. (*Also see* GROUNDSPEED.)

Indicated airspeed (IAS)—a measurement of the speed of an aircraft in relation to the air through which it is passing, as shown on a dial.

airspeed *continued*

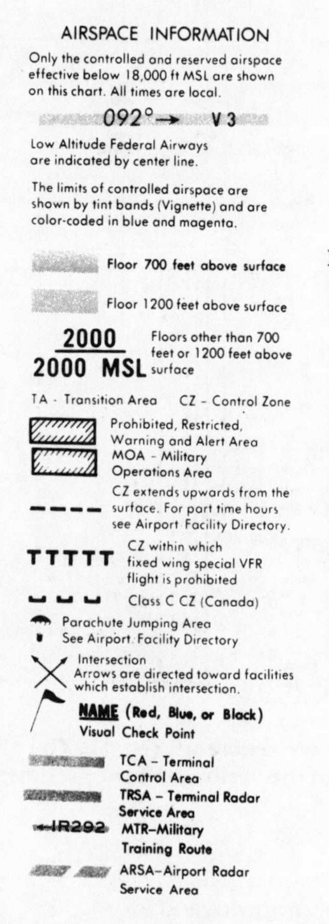
AIRSPACE INFORMATION
Only the controlled and reserved airspace effective below 18,000 ft MSL are shown on this chart. All times are local.
092° V 3
Low Altitude Federal Airways are indicated by center line.
The limits of controlled airspace are shown by tint bands (Vignette) and are color-coded in blue and magenta.
Floor 700 feet above surface
Floor 1200 feet above surface
2000
2000 MSL
Floors other than 700 feet or 1200 feet above surface
TA - Transition Area
CZ - Control Zone
Prohibited, Restricted, Warning and Alert Area
MOA - Military Operations Area
CZ extends upwards from the surface. For part time hours see Airport Facility Directory.
TTTTT
CZ within which fixed wing special VFR flight is prohibited
Class C CZ (Canada)
Parachute Jumping Area
See Airport Facility Directory
Intersection
Arrows are directed toward facilities which establish intersection.
NAME (Red, Blue, or Black)
Visual Check Point
TCA - Terminal Control Area
TRSA - Terminal Radar Service Area
IR292
MTR-Military Training Route
ARSA-Airport Radar Service Area

Figure A-8. Airspace information.

Calibrated airspeed (CAS)—indicated airspeed corrected for individual instrument error and installation error. Note the airspeed correction table for your specific aircraft in its operating manual.

Equivalent airspeed (EAS)—calibrated airspeed which has also been corrected for a compressibility factor. Important to pilots operating aircraft at speeds above 250 knots and altitudes higher than 10,000 feet.

True airspeed (TAS)—since the standard airspeed indicator is calibrated to read at standard sea level conditions (that is, barometric pressure at 29.92 inches and temperature at 59°F.), to know his true airspeed, the pilot must make allowances for the temperature and atmospheric pressure at the altitude he's flying. By noting the outside temperature on the gauge in his cockpit, his calibrated airspeed, and his pressure altitude, he can compute his true airspeed.

airspeed indicator—An instrument showing the approximate speed of the craft through the air.

airstart—The starting of an aircraft engine while the aircraft is airborne, preceded by engine shutdown during training flights or by actual engine failure.

air taxi—Used to describe a helicopter/VTOL aircraft movement conducted above the surface but normally not above 100 feet AGL. The aircraft may proceed either via hover taxi or flight at speeds more than 20 knots. The pilot is solely responsible for selecting a safe airspeed/altitude for the operation being conducted.

air traffic—Aircraft operating in the air or on an airport surface, exclusive of loading ramps and parking areas.

air traffic control (ATC)—A service provided for the purpose of promoting the safe, orderly, and expeditious flow of air traffic; it includes airport, approach, and en route air traffic control service.

air traffic control clearance (ATC CLEARANCE)—An authorization by air traffic control for an aircraft to proceed under specified traffic conditions within controlled airspace for the purpose of preventing collision between known aircraft.

air traffic control radar beacon system (ATCRBS)—This system, sometimes referred to as secondary surveillance radar, consists of three main components:

1. *Interrogator*—Primary radar relies on a signal being transmitted from the radar antenna site and being reflected, or "bounced back," from an object (such as an aircraft). This reflected signal is then displayed as a "target" on the controller's radarscope. In the ATCRBS the interrogator, a ground-based radar beacon transmitter-receiver, scans in synchronism with the primary radar and transmits discrete

air traffic control radar beacon system (ATCRBS) ***continued***

radio signals which repetitiously request all transponders on the mode being used to reply. The replies received are then mixed with the primary returns and both are displayed on the same radarscope.

2. *Transponder*—This airborne radar beacon transmitter-receiver automatically receives the signals from the interrogator and selectively replies with a specific pulse group (code) only to those interrogations being received on the mode to which it is set. These replies are independent of, and much stronger than, a primary radar return.

3. *Radarscope*—The radarscope used by the controller displays returns from both the primary radar system and the ATCRBS. These returns, called targets, are what the controller refers to in the control and separation of traffic.

air traffic control radio frequency—118.0 through 121.4 MHz and other frequencies as indicated on sectional charts for appropriate airport towers.

air traffic controller—The Federal Aviation Administration employee responsible for directing movement of planes either on the ground or in the air at or near an airport which has a control tower.

air turbulence, clear (CAT)—Turbulence experienced by an aircraft in airspace without clouds. CAT can be caused by convective currents, obstructions to the wind flow, or sudden changes in wind speed or direction.

airway—A control area or portion thereof, established in the form of a corridor, the centerline of which is defined by radio navigational aids.

airway beacons—Used before today's electronic airway system. Rotating beacons spaced 10 to 15 miles apart along the airways between airports formed a "lighted" pathway with Morse-coded flashes corresponding to sites marked on the sectional map. A few still remain in remote mountain locations.

airworthy—A craft in condition suitable for safe flight.

alcohol—Extensive research has provided a number of facts about the hazards of alcohol consumption and flying. As little as 1 ounce of liquor, one bottle of beer, or 4 ounces of wine can impair flying skills, with the alcohol consumed in these drinks being detectable in the breath and blood for at least 3 hours. Even after the body completely destroys a moderate amount of alcohol, a pilot can still be severely impaired for many hours by hangover. There is simply no way of increasing the destruction of alcohol or alleviating a hangover. Alcohol also renders a pilot much more susceptible to disorientation and hypoxia.

A consistently high alcohol-related fatal aircraft accident rate serves to emphasize that alcohol and flying are a potentially lethal combination. The FARs prohibit pilots from performing crewmember duties within 8 hours after drinking any alcoholic beverage or while under the influence of alcohol. However, due to the slow destruction of alcohol, a pilot may stil be under influence 8 hours after drinking a moderate amount of alcohol. Therefore, an excellent rule is to allow at least 12 to 24 hours between "bottle and throttle," depending on the amount of alcoholic beverage consumed.

alert area—Any area indicated on aeronautical charts to inform the private pilot that a high volume of pilot training or other unusual aerial activity takes place there, so that he should be especially alert.

alf—Aloft.

alphabet, phonetic—For maximum clarity in transmission, aircraft call letters are indicated by the following words:

A: Alfa
B: Bravo
C: Charlie
D: Delta
E: Echo
F: Foxtrot
G: Golf
H: Hotel
I: India
J: Juliett
K: Kilo (kee-lo)
L: Lima (lee-ma)
M: Mike
N: November
O: Oscar
P: Papa
Q: Quebec
R: Romeo
S: Sierra
T: Tango
U: Uniform
V: Victor
W: Whiskey
X: Xray
Y: Yankee
Z: Zulu

alt—Altitude.

altering course—When a pilot is lost, it is dangerous to alter course. He should first try to determine his position by searching out landmarks and checking his charts, and if these are unproductive, he should contact the nearest ground station and confess his predicament or use the emergency frequency of 121.5 to seek help.

alternate airport—When filing a flight plan for his destination, a pilot is required to select an alternate airport nearby for landing should weather or other conditions at his estimated time of arrival prohibit his landing at his original destination.

altimeter—An instrument that measures the relative altitude of the aircraft by measuring atmospheric pressure.

altimeter errors—Most pressure altimeters have some mechanical or installation errors. The deviation from normal is posted on a scale in the cockpit and should be taken into account when any altimeter settings are made. The importance of frequently obtaining current altimeter settings cannot be overemphasized. If you do not reset your altimeter when flying from an area of high pressure or high temperature into an area of low temperature or low pressure, your aircraft will be closer to the surface than the altimeter indicates. An inch error on the altimeter equals 1,000 feet of altitude. To quote an old saw: "Going from a high to a low (or a hot to a cold), look out below." The reverse situation: without resetting the altimeter when going from a low temperature or low pressure area into a high temperature or high pressure area, the aircraft will be higher than the altimeter indicates.

altimeter setting—The normal setting at sea level is 29.92, but pilots should set their altimeters to the setting at the airport from which they depart, as given them from stations en route, or as shown on their charts.

altimetry—Experience gained with the present volume of operations has led to the adoption of a standard altimeter setting for flights operating at 18,000 feet MSL or above. (*See* AIRMAN'S INFORMATION MANUAL.)

altitude—The height of a level, point, or object measured in feet Above Ground Level (AGL) or from Mean Sea Level (MSL). (*See* FLIGHT LEVEL.)

altitude—The five types are detailed below:

Indicated altitude—assumed height of aircraft above mean sea level (MSL), as read directly from the altimeter without correction for instrument error and uncompensated for variation from standard atmospheric conditions.

Pressure altitude—altitude indication when the altimeter setting window has been adjusted to 29.92 inches barometric pressure.

Density altitude—pressure altitude corrected for temperature variations.

True altitude—the actual height of the aircraft above sea level.

Absolute altitude—the height of the aircraft above ground level (AGL).

altitude readout/automatic altitude report—An aircraft's altitude, transmitted via the Mode C transponder feature, that is visually displayed in 100-foot increments on a radar scope having readout capability.

altitude reservation (ALTRV)—Airspace utilization under prescribed conditions, normally employed for the mass movement of aircraft or for other special user requirements which cannot otherwise be accomplished. ALTRVs are approved by the appropriate Federal Aviation Administration facility.

altitude restriction—An altitude or altitudes stated in the order flown which are to be maintained until reaching a specific point or time. Altitude restrictions may be issued by ATC due to traffic, terrain, or other airspace considerations.

altitude restrictions are canceled—Controller's expression meaning adherence to previously imposed altitude restrictions is no longer required during a climb or descent.

altitudes and aircraft radio reception distance—Assuming zero elevation of the radio transmitter, an aircraft that is 1,000 feet in the air can receive a station about 45 miles away. At 5,000 feet it can normally receive a station 100 miles away.

altitudes—(*Also see* MINIMUM SAFE ALTITUDE; VFR CRUISING ALTITUDES/FLIGHT LEVELS.)

alto—Prefix to cloud formations meaning "middle" height.

alto cumulus—Puffy, billowy clouds, usually found between 6,500 and 16,500 feet.

alto stratus—Layered horizontal clouds between 6,500 and 16,500 feet.

American Air Almanac—Lists the times of official sunrise and sunset in different parts of the country, which the FAA uses to regulate pilots for night flight experience.

amdt—Amendment.

ams—Air mass.

A.M. weather—(*See* WEATHER, A.M.)

anemometer—A meteorological device for measuring the velocity of wind.

angle of attack—The angle of the wing in relation to oncoming air in an aircraft's flight path.

angle of bank—The degree a plane is tilted from horizontal in making a turn.

angle of climb, best—This allows the plane to gain the greatest altitude in the least horizontal distance. It is made at an airspeed about 25 percent above stall speed. It is useful when there is an obstruction at the end of the field which must be cleared. This angle of climb should not be prolonged or it will cause the engine to overheat.

angle of incidence—The angle formed by the longitudinal axis of the airplane and the chord line of the wing (the angle at which the wing is mounted onto the fuselage).

AOPA—Aircraft Owners and Pilots Association, 421 Aviation Way, Frederick, MD 21701. Membership details are available on request.

apch—Approach.

apchg—Approaching.

App Con—Approach control.

approach control service—Air traffic control service, provided by a terminal area traffic control facility, for arriving and departing VFR and IFR aircraft and on occasion en route aircraft.

approach controller—The Federal Aviation Administration employee who provides the incoming aircraft with heading information to place it on the final approach course before it is handed over to the tower controller for landing instructions.

approach fix—The fix from or over which final approach (IFR) to an airport is executed.

approach gate—That point on the final approach course which is 1 mile from the approach fix on the side away from the airport or 5 miles from the landing threshold, whichever is farther from the landing threshold.

approach light system (ALS)—An airport lighting facility which provides visual guidance to landing aircraft by radiating light beams in a directional pattern by which the pilot aligns the aircraft with the extended centerline of the runway on his final approach for landing. Condenser-Discharge Sequential Flashing Lights/Sequenced Flashing Lights may be installed in conjunction with the ALS at some airports.

approach, power—The power approach is used in light airplanes under two different operating conditions: (1) for short field approaches (especially over obstructions) and (2) for landing in turbulent air to permit immediate adjustments of the approach slope when updrafts or downdrafts are present.

approach procedures—All airports (unless specifically noted otherwise because of nearby obstructions, multiple use of airspace, and so on) use a left-hand (counterclockwise) pattern approach to the landing runway. On an airport with dual runways of the same heading, the left runway uses a left pattern, the right runway a right pattern (*See* Figure A-9.)

approach sequence—The order in which aircraft are positioned while awaiting approach clearance or while on approach.

approach speed—The recommended speed contained in aircraft manuals used by pilots when making an approach to landing. This speed will vary for different segments of an approach as well as for aircraft weight and configuration.

approach to stall—When the plane is allowed to develop an excessive nose-high attitude, by back pressure being applied to the con-

approach to stall *continued*

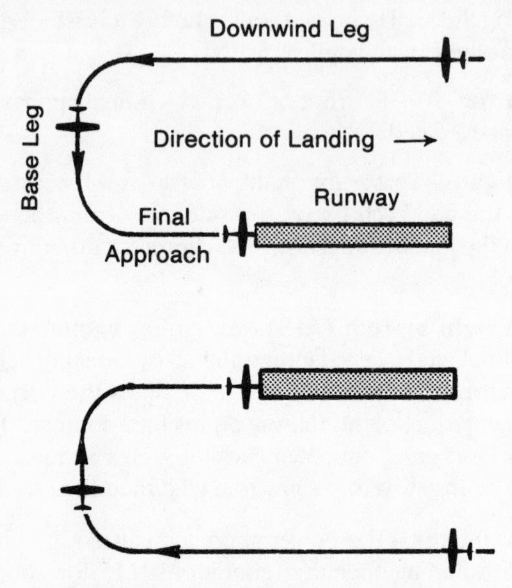

Figure A-9. Approach procedures.

trols, speed decreases until the plane starts to shudder prior to stalling. Recovery is accomplished by lowering the nose, applying full power, and allowing airspeed to build up to cruise speed before reducing throttle power.

apron/ramp—A defined area, on a land airport, intended to accommodate aircraft for purposes of loading or unloading passengers or cargo, refueling, parking or maintenance. With regard to seaplanes, a ramp is used for access to the apron from the water.

aprxly—Approximately.

ARC—The track over the ground of an aircraft flying at a constant distance from a NAVAID by reference to distance measuring equipment.

area forecast (FA)—This forecast covers flying weather expected in an 18-hour period (with an outlook on an additional 12 hours) and usually for parts of two or more states.

area navigation (RNAV) route system—Area navigation systems permit aircraft operations on any desired course within the coverage of station-referenced navigation signals or within the limits of self-contained system capability. Typical of such systems are VOR/DME RNAV, INS, LORAN C, and OMEGA.

Except for a few high altitude Alaskan routes, there is no longer a U.S. RNAV route system. Provisions are retained for designation of RNAV routes if they are required; however, these routes would be depicted on En Route High Altitude Charts. Special certification requirements for RNAV equipment apply to these routes.

In the absence of established RNAV routes, random RNAV routes may be filed, but radar monitoring by ATC will be required.

arnd—Around.

arpt—Airport.

arrival—(*See* VFR FLIGHT PLAN.)

ARSA—(*See* AIRPORT RADAR SERVICE AREA.)

ARSR—Air route surveillance radar.

ARTCC—Air route traffic control center.

artificial horizon—An instrument whereby the horizontal bar of a *T* remains fixed on the horizon despite the position of the aircraft. In haze and clouds it helps the pilot maneuver the plane to correct flight position.

ARTS—Automated radar terminal system.

ASDE—Airport surface detection equipment.

ASL—Above sea level.

ASR—Airport surveillance radar.

ATC—Air trafffic control.

ATC advises—Used to prefix a message of noncontrol information when it is relayed to an aircraft by other than an air traffic controller.

ATC clearances—When air traffic clearance has been obtained under either the visual or instrument flight rules, the pilot in command of the aircraft must not deviate from the provisions thereof unless an amended clearance is obtained.

ATC clears—Used to prefix an ATC clearance when it is relayed by other than an air traffic controller.

ATC requests—Used to prefix an ATC request when it is relayed to an aircraft by other than an air traffic controller.

ATCRBS—Air traffic control radar beacon system.

ATCT—Air traffic control tower.

ATIS—(*See* AUTOMATIC TERMINAL INFORMATION SERVICE.)

atmosphere—The body of air surrounding the earth. It is composed mainly of the gases nitrogen and oxygen.

atmospheric pressure—The force exerted by the weight of the atmosphere. At sea level this is 14.7 pounds per square inch. At sea level the atmospheric pressure in a barometer causes a column of mercury to rise 29.92 inches into an evacuated tube. The *actual* pressure depends on altitude, temperature, and density of the air.

attitude—The position of an airplane considering the inclination of its axes in relation to the horizon.

attitude indicator—An instrument showing the plane's position in relation to the natural horizon. (*Also see* ARTIFICIAL HORIZON.)

automated radar terminal system (ARTS)—Equipment designed to provide controllers with an alphanumeric display of aircraft identification and ground speed on aircraft equipped with transponders, along with altitude readout on those aircraft capable of automatic altitude reporting (mode C). This information is displayed on the controller's radar display as a data block and automatically tracks those aircraft.

auto—Automatic.

automatic altitude reporting—That function of a transponder which responds to Mode C interrogations by transmitting the aircraft's altitude in 100-foot increments.

automatic direction finder (ADF)—Sometimes called the radio compass, this instrument has an azimuth dial with the 0 or 360 degrees representing the nose of the aircraft. When the pilot tunes the receiver to the frequency of the nearest low or medium frequency radio transmitter, a pointer automatically shows him its position relative to his aircraft. For example, if he tuned into a commercial broadcast station at 700 kc and the pointer settles at 60 degrees, he knows the station is 60 degrees off to the right of his aircraft.

automatic pilot—A gyroscopic device for operating the flight controls without assistance from the pilot. It is commonly installed in large airplanes used for flights of considerable duration.

aux—Auxiliary.

auxiliary lights—(*See* LIGHTS, AUXILIARY.)

avbl—available.

automatic terminal information service (ATIS)—The continuous broadcast of recorded noncontrol information in selected terminal areas. Its purpose is to improve controller effectiveness and to relieve frequency congestion by automating the repetitive transmission of essential but routine information, e.g. "Los Angeles information Alpha One three zero zero Greenwich Weather, measured ceiling two thousand overcast, visibility three, haze, smoke, temperature seven one, dewpoint five seven, wind two five zero at five, altimeter two niner niner six I-L-S Runway Two Five Left approach in use. Runway Two Five Right closed, advise you have Alpha." (*See* AIM.)

AWOS—Automated weather observing system automatically gathers weather data from various locations around an airport and transmits to pilots through computer-generated voice messages over a radio frequency of a local NAVAID. Data provided include wind direction and speed, temperature, dewpoint, visibility, sky conditions and ceiling, precipitation and altimeter setting. AWOSs are receivable at 3,000 feet AGL to at least 25 nautical miles of the AWOS site. Check sites and frequencies in the *Airport/Facility Directory.*

awy—Airway.

axes of an airplane—The three theoretical lines which intersect at the center of gravity of the airplane: the longitudinal axis extending from nose to tail, about which the plane rolls as controlled by the ailerons; the lateral axis extending horizontally from wing tip to wing tip, from which the plane pitches, as controlled by the elevators; the vertical axis, extending through the airplane from top to bottom, from which the plane yaws, as controlled by the rudder.

azimuth—The distance in degrees in a clockwise direction from the 360 or 0 degrees north point.

B

B in the phonetic alphabet is Bravo (pronounced bra-voh).

bail out—To jump from an airplane in flight, using a parachute.

balance—The proper loading of passengers and baggage for the safest and most efficient operation of the aircraft.

ball—(*See* TURN AND SLIP INDICATOR.)

balloons, avoid flight beneath unmanned—The majority of unmanned free balloons currently being operated have extending below them either a suspension device to which the payload or instrument package is attached, or a trailing wire antenna, or both. In many instances these balloon subsystems may be invisible to the pilot until his aircraft is close to the balloon, thereby creating a potentially dangerous situation. Therefore, good judgment on the part of the pilot dictates that aircraft should remain well clear of all unmanned free balloons and flight below them should be avoided.

Pilots are urged to report any unmanned free balloons sighted to the nearest FAA ground facility with which communication is established. Such information will assist FAA ATC facilities to identify and flight follow unmanned free balloons operating in the airspace.

bank—To roll about the longitudinal axis of the airplane as in a turn. Banking is controlled with the ailerons.

barometric pressure—A difference of .1 inch can equal 100 feet on the altimeter; a 1-inch difference equals 1,000 feet. Be sure you know the correct altimeter setting for the terrain over which you are flying. If you fly from a high pressure area into a low pressure area without resetting your altimeter, you will actually be *lower* than the indicated altitude. This can be dangerous when reading charts for obstructions, and so on.

base leg—(*See* TRAFFIC PATTERN.)

basic flight information—*Airman's Information Manual.* Issued every 112 days (by subscription) from the Superintendent of Documents, United States Government Printing Office, Washington, D.C. 20402.

basic VFR minimum weather conditions—(*See* VFR FLIGHT WEATHER MINIMUMS IN CONTROLLED AIRSPACE; VFR FLIGHT WEATHER MINIMUMS IN UNCONTROLLED AIRSPACE.)

BCM—Back course marker.

bcn—Beacon.

bcst—Broadcast.

BD—Blowing dust.

beacons, aeronautical—(*See* AIRPORT ROTATING BEACONS.)

beacons, airway—(*See* AIRWAY BEACONS.)

beacons, daylight operation—(*See* DAYLIGHT BEACON OPERATION.)

beacons, locator—(*See* LOCATOR BEACONS.)

beacons, marker—(*See* MARKER BEACONS.)

beacons, operation of aircraft rotating—(*See* OPERATION OF AIRCRAFT ROTATING BEACONS.)

beacons, rotating—(*See* AIRPORT ROTATING BEACONS.)

bearing—The horizontal direction to or from any point, usually measured clockwise from true north, magnetic north or some other reference point, through 360 degrees. (*See* NONDIRECTIONAL RADIO BEACON.)

Beaufort scale—Developed in the early 19th century by an admiral of the British navy to estimate wind speeds from their effect on sails. The scale was subsequently adapted for land use.

below minimums—Weather conditions below the minimums prescribed by regulation for the particular action involved, e.g., landing minimums, take-off minimums.

bends, the—(*See* SCUBA DIVING.)

Bernoulli's theory—The Italian scientist Daniel B. Bernoulli (bĕr-nōōl'ē, 1700–1782) discovered that if the velocity of a fluid in a stream tube is increased at a particular point, the pressure will decrease at that point. This theory as applied to the wings of aircraft is that if you increase the velocity of the air over the top of the wing (by the design of the wing), then the pressure is decreased at that point. The corresponding increased pressure *under* the wing at that point causes lift.

best angle-of-climb airspeed—The airspeed which results in the aircraft achieving the greatest altitude in the least distance (usually about 1.25 times the stall speed of the individual aircraft).

best rate-of-climb airspeed—The airspeed which produces the greatest altitude in the least time (usually about 1.5 times the stall speed).

bfdk—Before dark.

bienniel flight review—No person may act as pilot in command of an aircraft unless, within the preceding 24 months, he has:

1. Accomplished a flight review given to him, in an aircraft for which he is rated, by an appropriately certificated instructor or other person designated by the Administrator; and
2. Had his log book endorsed by the person who gave him the review certifying that he has satisfactorily accomplished the flight review.

However, a person who has, within the preceding 24 months, satisfactorily completed a pilot proficiency check conducted by the FAA, an approved pilot check airman, or a U.S. armed force for a pilot certificate, rating, or operating privilege, need not accomplish the flight review.

biplane—An airplane with two wings, one above the other.

bird hazards—All pilots are requested to contact the nearest tower or flight service station when they observe large flocks of birds and report the location, bird species, if known, quantity, altitude, and direction of flight. Known migratory patterns and dates of flight of certain species are published in *Advisory Circular* AC-150/6200-3.

Notices of airports which have been reported to the FAA as having high density bird populations are published in the *Notices to Airmen* (Class II) publication until they can be transferred to the remarks section of the *Airport/Facility Directory.*

Since migrating waterfowl tend to dive when closely approached by aircraft, pilots are warned not to fly directly under migrating flocks of swans, geese, or ducks.

bird strike/incident, reporting—Pilots involved in a bird strike or incident are urged to make a report using a special form available at any Federal Aviation Administration area office, General Aviation district office, flight service station, and Air Carrier district office. The data and information derived from these reports will be used to develop standards to cope with this expensive hazard to aircraft and for habitat control methods on or adjacent to airports.

biscuit gun—The high intensity light used by control tower operators to give landing instructions to pilots whose radios have failed.

bkn—Broken.

bldg—Building.

blind spot/blind zone—An area from which radio transmissions and/or radar echoes cannot be received. The term is also used to describe portions of the airport not visible from the control tower.

blo—Below.

blue tinted bands—On aeronautical charts they indicate controlled airspace with a floor 1,200 feet above surface.

blzd—Blizzard.

BN—Symbol for blowing sand.

bndry—Boundary.

boost—Used to denote an increase in manifold pressure or throttle setting.

booster—An electrical device to aid in starting an airplane engine or a power device for aiding the movement of flight controls in heavy aircraft.

braking action (good, fair, poor, or nil)—A report of conditions on the airport movement area providing a pilot with a degree/quality of braking that he might expect. Braking action is reported in terms of good, fair, poor, or nil.

braking action advisories—When tower controllers have received runway braking action reports which include the terms "poor" or "nil," or whenever weather conditions are conducive to deteriorating or rapidly changing runway braking conditions, the tower will include on the ATIS broadcast the statement, "BRAKING ACTION ADVISORIES ARE IN EFFECT." During the time Braking Action Advisories are in effect, ATC will issue the latest braking action report for the runway in use to each arriving and departing aircraft. Pilots should be prepared for deteriorating braking conditions and should request current runway condition information if not volunteered by controllers. Pilots should also be prepared to provide a descriptive runway condition report to controllers after landing.

brf—Brief.

brg—Bearing.

BS—Symbol for blowing snow.

btn—Between.

btr—Better.

buffeting—The beating effect on the structure of an aircraft when there is improper airflow over the wings.

bumpy air—Unequal heating of ground masses (trees vs. desert, for example) causes considerable difference in the vertical flow of air up from the ground and sometimes results in rough air as the plane moves through the upward draft of air.

burble—An increase in the angle of attack of an aircraft beyond the stalling angle (approximately 20 degrees) when air is no longer flowing properly over the wing surface. It creates a turbulent condition in the air above the upper surface of the wing, called a burble. When this breaking away (burbling) of the air occurs, the plane will shudder and shake and then lose lift immediately. Antidote: push forward on the elevator control to reduce the angle of attack; apply full power until normal cruising speed is attained.

byd—Beyond.

C

C in the phonetic alphabet is Charlie (pronounced char-lee).

C—Abbreviation for Celsius (or centigrade).

C—symbol for continental air mass.

C formula—Formula for converting a Celsius (or centigrade) temperature to Fahrenheit: $C° = 5/9(F° - 32)$.

calibrated airspeed—(*See* AIRSPEED.)

call-up—Initial voice contact between a facility and an aircraft, using the identification of the unit being called and the unit initiating the call.

Canadian airspace, operating in—Pilots planning flights in Canadian airspace are cautioned to review Canadian Air Regulations. Special attention should be given to those Canadian regulations which differ from United States Federal Aviation Regulations. Prior to flight within Canadian airspace, visiting pilots are urged to obtain and study the Canadian publication *Air Tourist Information.* The *Air Tourist Information* booklet is available from Transport Canada, Aeronautical Information Services, (AISP/A), Ottawa, Ontario, Canada K1A 0N8. Pilots should note in the *Air Tourist Information* publication the procedures which differ from those in effect in United States airspace in particular:

1. "1000 on top" is IFR flight, that is, "1000 on top" is not permitted in Canadian airspace under the visual flight rules (VFR).
2. The regulations concerning VFR flight within Class B airspace, that is, controlled VFR.
3. The altimeter setting procedures within the altimeter setting region and within standard pressure region.
4. Equipment requirements for flights within sparsely settled areas.

carbon monoxide—A colorless, odorless, tasteless gas, a product of an internal cumbustion engine which is always present in exhaust fumes. Even minute quantities of carbon monoxide breathed over a long period of time may lead to dire consequences. For biochemical reasons, carbon monoxide has a greater ability to combine with the hemoglobin of the blood than oxygen. Furthermore, once carbon monoxide is absorbed in the blood, it sticks like glue to the hemoglobin and actually prevents the oxygen from attaching to the hemoglobin. Most heaters in light aircraft work by air flowing over the manifold. So if you have to use the heater, be wary if you smell exhaust fumes. The onset of symptoms is insidious, with "blurred" thinking, a possible feeling of uneasiness, and subsequent dizziness. Later, headache occurs. Immediately shut off the heater, open the air ventilators, descend to lower altitudes, and land at the nearest airfield. Consult an aviation medical examiner. It may take several days to fully recover and clear the body of the carbon monoxide.

carburetor heat—Used to forestall icing in the carburetor. Carburetor heat is usually applied about 30 seconds before reducing power and under conditions mentioned in carburetor icing.

carburetor icing—A frequent cause of engine failure and indicated by a loss of r.p.m. and a decrease in manifold pressure. Carburetor icing is most likely to occur when the temperature is between 49 and 60 degrees F. with visible moisture or high humidity. If the symptoms described above occur, the carburetor heat should be turned full-on immediately.

cardinal altitudes or flight levels—"Odd" or "even" thousand-foot altitudes or flight levels. Examples: 5,000, 6,000, 7,000, FL 250, FL 260, FL 270.

CAS—Calibrated Airspeed. (*See* AIRSPEED.)

CAT—Clear air turbulence.

caution area—Airspace within which military activities are conducted that are not hazardous but are of interest to nonparticipating pilots.

CAVOK—Cloud and visibility OK.

ceiling (in aircraft)—The maximum altitude the airplane is capable of obtaining under standard conditions.

ceiling (in meteorology)—The height above the earth's surface of the lowest layer of clouds or obscuring phenomena that is reported as "broken," "overcast," or "obscuration" and not classified as "thin" or "partial." The ceiling is classified in several ways and shown by a letter in weather reports preceding the ceiling height. Some of the more important of these letters are M (measured), E (estimated), W (indefinite). If one of these letters does not precede the cloud symbol or if thin broken, or overcast clouds exist, there is no official ceiling. The letter V immediately following a height indicates a variable ceiling.

Celsius—A thermometric scale in which the interval between the freezing point (0 degrees) and boiling point is 100 degrees.

center—(*See* AIR ROUTE TRAFFIC CONTROL CENTER.)

center of gravity—That point in an aircraft in which, when power is on, all forces of weight, lift, thrust, and drag are in balance and level flight is maintained.

center of pressure—That point in the airfoil section of the wing through which all the upward forces of lift are considered to act for the purpose of computations considering the airplane as a whole.

centigrade—Same as Celsius.

central altitude reservation facility (CARF)—An air traffic service facility established to conduct the volume of coordination, planning, and approval of special user requirements under the altitude reservation concept.

cetificates—(*See* AIRCRAFT CERTIFICATES, MEDICAL CERTIFICATES, PILOT CERTIFICATE.)

CH—Compass heading.

chaff—Tiny strips of metal foil used to help a radar controller to detect a lost plane. If the pilot equipped with chaff is unable to communicate by radio, he should: (1) fly a straight course, opening and dropping a box of chaff every 2 miles until four drops have been made; (2) continue on course for 2 miles and then make a 360 degree turn, at a rate of 3 degrees per second, to the left; and (3) repeat the 360 degree turn at 2-mile intervals until four turns have been completed. The chaff pattern will enhance the possibility of detection by the radar controller and will assist him in providing assistance. (*See* Figure C-1.)

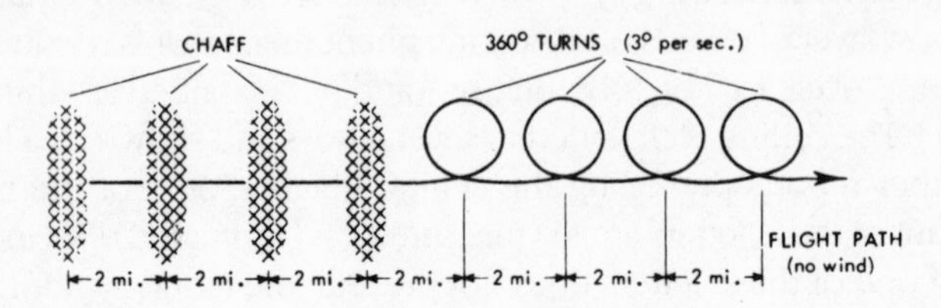

Figure C-1. Chaff pattern.

chandelle—A training maneuver which requires a high degree of planning and results in a coordinated climb and turn with a 180 degree change in direction of the aircraft.

chg—Change.

charts—(*See* AERONAUTICAL CHARTS.)

check list—A reminder list for the pilot of items requiring his attention for preflight and flight operations. Most owners' manuals contain recommended checklists for the particular type of aircraft. By using these lists for every flight, the possibility of overlooking important items will be lessened.

checkpoint—A prominent landmark, either visual or radio, used in air navigation to establish the position of the aircraft in flight.

chinook—(*See* FOEHN.)

chord—A straight line joining the two points on a curve. In an airfoil, a chord is an imaginary line from the center of the leading edge to the trailing edge.

CIG—Abbreviation for ceiling. Also used to indicate marginal VFR conditions due to both ceiling and visibility. If winds of 25 knots or greater are forecast, the word "wind" is also included.

circle to runway (runway numbered)—Used by ATC to inform the pilot that he must circle to land because the runway in use is other than the runway aligned with the instrument approach procedure. When the direction of the circling maneuver in relation to the airport/runways is required, the controller will state the direction (eight cardinal compass points) and specify a left or right downwind or base leg as appropriate, e.g., "Cleared VOR Runway three six approach circle to Runway two two" or "Circle northwest of the airport for a right downwind to Runway two two."

circuit breaker—A device which takes the place of a fuse in breaking an electrical circuit in case of an overload. If the overload was temporary, breakers can be reset in most aircraft.

cirrus—(*See* CLOUDS.)

civil air regulations—Established by the United States government for safe operation of an aircraft by a competent pilot, now called Federal Aviation Regulations.

civil use of military fields—United States Army, Air Force, Navy, and Coast Guard fields are open to civil fliers only in emergency or with prior permission.

clear air turbulence (CAT)—Turbulence encountered in air where no clouds are present; more popularly applied to high-level turbulence associated with wind shear; often encountered in the vicinity of the jet stream. Clear air turbulence has become a very serious operational factor to flight operations at all levels and especially to jet traffic flying above 15,000 feet. The best available information on this phenomenon must come from pilots via the PIREPS procedures. (*See* WIND SHEAR.)

clearance limit—The fix, point, or location to which an aircraft is cleared when issued an air traffic clearance.

clearances—(*See* SPECIAL VFR CLEARANCES.)

cleared as filed—Means the aircraft is cleared to proceed in accordance with the route of flight filed in the flight plan. This clearance does not include the altitude, SID, or SID Transition.

cleared for takeoff—ATC authorization for an aircraft to depart. It is predicated on known traffic and known physical airport conditions.

cleared for the option—ATC authorization for an aircraft to make a touch-and-go, low approach, missed approach, stop and go, or full stop landing at the discretion of the pilot. It is normally used in training so that an instructor can evaluate a student's performance under changing situations. (*See* OPTION APPROACH.) (*See* AIM.)

cleared through—ATC authorization for an aircraft to make intermediate stops at specified airports without refiling a flight plan while en route to the clearance limit.

cleared to land—ATC authorization for an aircraft to land. It is predicated on known traffic and known physical airport conditions.

clearing an obstruction—It is sometimes required, when the obstruction is close to a landing field, to use a "best angle of climb" for departure and a slip or a power approach for landing. (*See* BEST ANGLE-OF-CLIMB AIRSPEED; POWER APPROACH; SLIP.)

clearing procedures, use of—Prior to taxiing onto a runway or landing area in preparation for takeoff, a pilot should scan the approach areas for possible landing traffic, executing appropriate clearing maneuvers to provide him a clear view of the approach areas. The same caution should be observed in climbs and descents and in the vicinity of VORs and airway intersections.

clearing the runway after landing—After landing, unless otherwise instructed by the control tower, aircraft should continue to taxi in the landing direction, proceed to the nearest turnoff, and exit the runway without delay. Do not make a 180 degree turn to taxi back on a landing runway without approval or specific instructions from the control tower to do so.

climb—Maneuver for gaining altitude. There are four types: normal climb, best angle climb, best rate of climb, and cruise climb. (*See* BEST ANGLE-OF-CLIMB AIRSPEED; BEST RATE-OF-CLIMB AIRSPEED; CRUISE CLIMB; NORMAL CLIMB.)

climb to VFR—ATC authorization for an aircraft to climb to VFR conditions within a control zone when the only weather limitation is restricted visibility. The aircraft must remain clear of clouds while climbing to VFR.

climbing turn—Once climb power and altitude are established, a roll into a bank sufficient to provide a standard rate turn left or right is made.

climbout—That portion of flight operation between takeoff and the initial cruising altitude.

closed runway or taxiway—(*See* RUNWAY MARKING.)

closed traffic—Successive operations involving takeoffs and landings or low approaches where the aircraft does not exit the traffic pattern.

closing the flight plan—(See FLIGHT PLAN, CLOSING, VFR/DVFR.)

cloud heights—In weather reports, cloud heights are given in feet above the ground at the terminal, not feet above sea level. Pilots usually report height values above mean sea level, since they determine heights by the altimeter. This is taken into account when disseminating and otherwise applying information received from pilots. ("Ceiling" heights are always above ground level.) In reports disseminated as PIREPS, height references are given the same as received from pilots, that is, above mean sea level (MSL or ASL).

clouds—The three basic forms are: stratus—flat streaks or layers; cumulus—puffy, billowy; and cirrus—thin, white, feathery shapes composed entirely of ice crystals at high altitudes (false cirrus—cirrus-like clouds at the summit of a thunder cloud, also called "thunderstorm cirrus.") (*See* Figures C-2, C-3, and C-4.)

Figure C-2. Stratus clouds.

Figure C-3. Cumulus clouds.

clouds, clearance from—(*See* VFR FLIGHT WEATHER MINIMUMS IN UNCONTROLLED AIRSPACE; VFR FLIGHT WEATHER MINIMUMS IN CONTROLLED AIRSPACE.)

clsd—Closed.

Co—County.

cockpit—The open space in the fuselage for pilot and passengers; in larger aircraft, just the pilot's compartment.

codes—The numbers assigned to the multiple pulse reply signals transmitted by ATCRBS and SIF transponders.

collision course correction procedure—Each aircraft alters its course to the *right.*

colors of on-the-ground and in-flight light signals—(*See* LIGHT SIGNALS.)

comlo—Compass locator.

Figure C-4. Cirrus clouds.

compass, gyro—A heading indicator which because of its gyroscopic properties is less affected by movement of the aircraft in turns, rough air, magnetic disturbance, etc. However, to maintain accuracy, since the gyro has a tendency to slowly move off the correct heading, it must be reset to the magnetic compass about every 15 minutes.

compass, magnetic—A heading indicator utilizing magnetized needles lined up with the magnetic field of the earth. If not subjected to local magnetic disturbances, it will indicate direction to the north magnetic pole.

compass errors—(*See* ACCELERATION ERROR, DEVIATION, MAGNETIC DIP, OSCILLATION ERROR, VARIATION.)

compass rose—A circle, graduated in degrees, printed on some charts or marked on the ground at an airport. It is used as a reference to either true or magnetic direction.

compensation or hire—(*See* PASSENGERS, CARRYING OF.)

compression—One of the four strokes of the engine: intake stroke, compression stroke, power stroke, exhaust stroke.

computer—A circular slide rule or an electronic device for aiding the pilot in navigation problems.

comsnd—Commissioned.

comsng—Commissioning.

CONSOLAN—A low frequency, long-distance NAVAID used principally for transoceanic navigation.

const—Construction.

cont—Continue/continuous/continuously.

contact—expression meaning:

1. Establish communications with (followed by the name of the facility and, if appropriate, the frequency to be used).

2. A flight condition wherein the pilot ascertains the attitude of his aircraft and navigates by visual reference to the surface.

conterminous United States—The 48 contiguous states and the District of Columbia.

continental control area—The Continental Control Area consists of the airspace of the 48 contiguous States, the District of Columbia, and Alaska, excluding the Alaska peninsula west of longitude 160 degrees 00 minutes 00 seconds W, at and above 14,500 feet MSL, but does not include:

1. The airspace less than 1,500 feet above the surface of the earth; or

2. Prohibited and Restricted areas, other than the Restricted areas listed in the FAR 71 Subpart D.

continental United States—The 49 states located on the continent of North America, and the District of Columbia.

control area—Airspace designated as Colored Federal Airways, VOR Federal Airways, control areas associated with jet routes outside the continental control area, additional control areas, control area extensions, and area low routes. Control areas do not include the continental control area, but unless otherwise designated, they do include the airspace between a segment of a main VOR Federal Airway and its associated alternate segments with the vertical extent of the area corresponding to the vertical extent of the related segment of the main airway. The vertical extent of the various categories of airspace contained in control areas is defined in FAR Part 71.

control sector—An airspace area of defined horizontal and vertical dimensions for which a controller, or group of controllers, has air traffic control responsibility, normally within an air route traffic control center or an approach control facility. Sectors are established based on predominant traffic flows, altitude strata, and controller workload. Pilot-communications during operations within a sector are normally maintained on discrete frequencies assigned to the sector. (*See* DISCRETE FREQUENCY.)

control surface—The hinged airfoils—ailerons, rudders, elevators—actuated by the pilot in controlling the attitude of the plane.

control tower—Towers have been established to provide for a safe, orderly, and expeditious flow of traffic on, and in the vicinity of, an airport. When the responsibility has been so delegated, towers also provide for the separation of IFR aircraft in the terminal areas. (*Also see* APPROACH CONTROL SERVICE.)

control zone—Controlled airspace which extends upward from the surface and terminates at the base of the continental control area. Control zones that do not underlie the continental control area have no upper limit. A control zone may include one or more airports and is normally a circular area within a radius of 5 statute miles and any extensions necessary to include instrument approach and departure paths. On sectional charts, the control zone is outlined by a broken blue line.

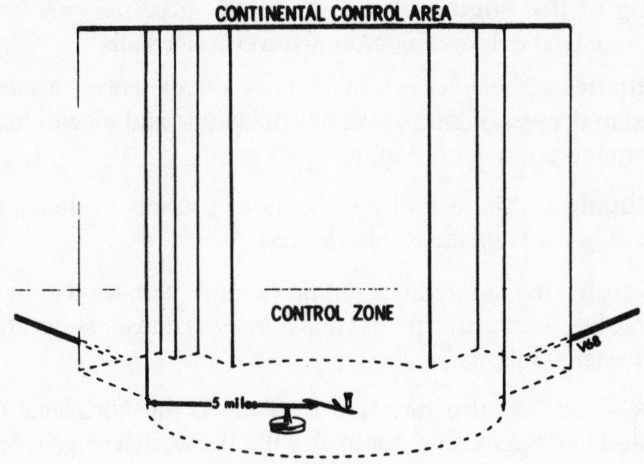

Figure C-5. Control zone.

controlled airspace—Airspace designated as continental control area, control area, control zone, or transition area, within which some or all aircraft may be subject to air traffic control. Controlled airspace is shown on aeronautical charts through color-coded blue and magenta tinted bands. It may be made up of an airport traffic area, continental control area, control area, control zone, or federal airway. (*See* the individual listings for the first four terms. *Also see* FEDERAL AIRWAYS, USE OF; VFR FLIGHT WEATHER MINIMUMS IN CONTROLLED/UNCONTROLLED AIRSPACE.)

controls—The elevators, ailerons, and rudders which affect the attitude of the plane in flight or taxiing.

convective sigmet/convective significant meteorological information—A weather-advisory concerning convective weather significant to the safety of all aircraft. Convective SIGMETs are issued for tornadoes, lines of thunderstorms, embedded thunderstorms of any intensity level, areas of thunderstorms greater than or equal to VIP level 4 with an areal coverage of 4/10 (40%) or more, and hail ¾ inch or greater. (*See* SIGMET, AIRMET.)

converging aircraft—Aircraft to the right of another has the right-of-way.

conversion scales—Located on the bottoms of some charts to show the comparison between kilometers, nautical miles, and statute miles.

cooling of the engine—In most general aircraft, engine cooling takes place by the flow of air around the cylinder walls.

coordinates—The intersection of lines of reference, usually expressed in degrees/minutes/seconds of latitude and longitude, used to determine position or location.

coordination—The use of two or more controls in their proper relationship to obtain the results desired.

correction—In voice radio communication, a term used to indicate "an error has been made in this (or a previous) transmission, and the correct version follows."

course—The intended direction of flight in the horizontal plane measured in degrees from north; also the ILS localizer signal pattern usually specified as front course or back course.

crab—Placing the aircraft in an attitude into the wind while still maintaining a correct forward flight path over the ground.

crash locator beacon—An electronic device attached to the aircraft structure as far aft as practicable in the fuselage, or in the tail surface, in such a manner that damage to the beacon will be minimized in the event of crash impact. It may be automatically ejectable or permanently mounted. If it is automatically ejectable, it will also have provision for manual removal and operation. The beacon operates from its own power source on 121.5 MHz and/or 243 MHz, preferably on both emergency frequencies, transmitting a distinctive downward swept audio tone for homing purposes, and is designed to function without human action after an accident.

crashed aircraft—If you observe a crashed aircraft, do the following: (1) determine if the crash is marked with a yellow cross; if so, the crash has already been reported and identified; (2) determine, if possible, the type and number of aircraft and whether there is evidence of survivors; (3) fix, as accurately as possible, the exact location of the crash with reference to a navigational aid or geographic description; (4) if circumstances permit, orbit the scene to guide in other assisting units or until relieved by another aircraft; (5) transmit information to the nearest Federal Aviation Administration or other appropriate radio facility; (6) immediately after landing, make a complete report to the nearest Federal Aviation Administration, Air Force, or Coast Guard installation. The report may be made by long-distance collect telephone.

critical flight situations—Takeoff and departure; approach to a landing; the performance of abrupt maneuvers at relatively slow airspeeds.

cross-country flight—Experience required for a private pilot's license of 10 hours solo cross-country flying to a landing place more than 50 nautical miles from the place of departure. One flight must be at least 300 nautical miles with landings at a minimum of three points, one of which is at least 100 nautical miles from the original departure point.

crosswind—

1. When used by a controller concerning the traffic pattern, the word means "crosswind leg." (*See* TRAFFIC PATTERN.)

crosswind *continued*

2. When used concerning wind conditions, the word means a wind not parallel to the runway or the path of an aircraft. (*See* CROSSWIND COMPONENT.)

crosswind component—The wind component measured in knots at 90 degrees to the longitudinal axis of the runway.

crosswind landing techniques—

1. The windward wing is lowered slightly and the flight path maintained by use of the rudder.

2. A heading into wind (crab) is maintained and then the flight path corrected by the rudder.

crosswind leg—(*See* TRAFFIC PATTERN.)

crosswind take-off technique—In single-engine aircraft the ailerons must be held into the wind and the flight path held straight with the rudder.

crs—course.

cruise—Used in an ATC clearance to authorize a pilot to conduct flight at any altitude from the minimum IFR altitude up to and including the altitude specified in the clearance. The pilot may level off at any intermediate altitude within this block of airspace. Climb/descent within the block is to be made at the discretion of the pilot. However, once the pilot starts descent and verbally reports leaving an altitude in the block he may not return to that altitude without additional ATC clearance. Further, it is approval for the pilot to proceed to and make an approach at destination airport.

cruise climb—Used instead of a normal climb for a long, low rate of climb, for more efficient fuel economy on cross-country trips.

cruise control—Operating the aircraft at maximum efficiency on an extended flight.

cruising altitude—A level determined by vertical measurement from mean sea level. (*See* VFR CRUISING ALTITUDES/FLIGHT LEVELS.)

cruising speed—Normal operating speed for the particular aircraft's maximum efficiency.

C's, the four—

1. Confess your predicament to any ground radio station or the emergency frequency 121.5. Do not wait too long. Give search and rescue a chance!

2. Communicate with your ground link and pass as much as possible of the distress message on first transmission.

3. Climb if possible for better radar and DF detection. If you are flying at low altitude, the chance for establishing radio contact is improved by climbing; also chances of alerting radar systems are sometimes improved by climbing. NOTE: Unauthorized climb or descent under IFR conditions within controlled airspace is not permitted except in emergency. Any variation in altitude, in connection with flying radar patterns, will be unknown to air traffic control except at radar locations having height-finding capabilities. Air traffic control will operate on the assumption that the provisions of Federal Aviation Regulations Part 91 are being followed by the pilot.

4. Comply—especially comply—with advice and instructions received, if you really want help. Assist the ground "communications control" station to control communications on the distress frequency on which you are working (as that is the distress frequency for your case). Tell interfering stations to maintain silence until you call. Cooperate!

csdrbl—Considerable.

CTAF (common traffic advisory frequency)—The key to communicating at an uncontrolled airport is selection of the correct common frequency. The contraction CTAF is synonymous with this program. A CTAF is a frequency designated for the purpose of carrying out airport advisory practices while operating to or from an uncontrolled airport. The CTAF may be a UNICOM, MULTICOM, FSS, or tower frequency and is identified in appropriate aeronautical publications.

The CTAF frequency for a particular airport can be obtained by contacting any FSS. Use of the appropriate CTAF, combined with a visual alertness and application of the AIM-recommended good operating practices, will enhance safety of flight into and out of all uncontrolled airports.

ctc—Contact.

cumulus—(*See* CLOUDS.)

cushioning effect—The temporary gain in lift during a landing, due to the compression of the air between the wings of an airplane and the ground.

customs regulations—(*See* FLIGHT OUTSIDE THE UNITED STATES.)

cvr—Cover.

D

D in the phonetic alphabet is Delta (dell-tah)

D—Weather report symbol for dust; also used to precede the figure for estimated height of cloud layer of cirriform clouds on the basis of persistency.

DADS—Digital air data system. DADS uses computers and solid-state components to replace electromechanical devices, thus providing airline crews with constant information on altitude, air speed, temperature, and related functions.

dalgt—Daylight.

daylight beacon operation—Operation of an airport rotating beacon during hours of daylight means that the ground visibility in the control zone is less than three miles and/or the ceiling is less than 1,000 feet and that a traffic clearance is required for landings, takeoffs, and flight in the traffic pattern.

dcmsnd—Decommissioned.

dcr—Decrease.

dead reckoning—Plotting a trip with reference only to charts and reference to visual landmarks as opposed to flying the VOR radio navigation system. [*See* VHF OMNIDIRECTIONAL RANGE (VOR).]

dead stick landing—Landing an aircraft without power.

declaring an emergency—The pilot should remember that he has four means of declaring an emergency: (1) emergency SQUAWK from transponders; (2) sending an emergency message to any nearby radio receiving station or 121.5 MHz, the emergency frequency; (3) flying triangular pattern; and (4) dropping chaff.

deepening—Decreasing pressure in the center of a low-pressure system.

defense VFR flight plan—During defense emergency or air defense emergency conditions, depending on instructions received from the military under the Security Control, Air Traffic and Air Navigation Aids (SCATANA) Plan, VFR flights may be directed to land at the nearest available airport, and IFR flights will be expected to proceed as directed by ATC. Pilots on the ground may be required to file a flight plan and obtain an approval (through Federal Aviation Administration) prior to conducting flight operation. Pilots should guard an ATC or FSS frequency on all flights.

defense visual flight rules (DVFR)—Rules applicable to flights within an ADIZ conducted under the visual flight rules in FAR Part 91.

degrees—The three digits of the magnetic course, bearing, heading, or wind direction. The word "degrees" should be added for wind direction, which is usually degrees magnetic. The word "true" must be added when it applies, such as for winds aloft or true course. Examples: (magnetic course) 005—ZERO ZERO FIVE; (true course) 050—ZERO FIVE ZERO TRUE; (magnetic bearing) 360—THREE SIX ZERO; (magnetic heading) 100—ONE ZERO ZERO; (wind direction) 215—TWO ONE FIVE DEGREES.

delay indefinite (reason if known) expect further clearance (time)—Used by ATC to inform a pilot when an accurate estimate of the delay time and the reason for the delay cannot immediately be determined; e.g., a disabled aircraft on the runway, terminal or center area saturation, weather below landing minimums, etc. (*See* EXPECTED FURTHER CLEARANCE TIME [EFC].)

density altitude—(*See* ALTITUDE.)

density altitude effects—Performance figures in the aircraft owner's handbook for length of take-off run, horsepower, rate of climb, etc., are generally based on standard atmosphere conditions (59 degrees Fahrenheit (15 degrees Celsius), pressure 29.92 inches of mercury) at sea level. However, inexperienced pilots, as well as experienced

density altitude effects ***continued***

pilots, may run into trouble when they encounter an altogether different set of conditions. This is particularly true in hot weather and at higher elevations. Aircraft operations at altitudes above sea level and at higher than standard temperatures are commonplace in mountainous areas. Such operations quite often result in a drastic reduction of aircraft performance capabilities because of the changing air density. Density altitude is a measure of air density. It is not to be confused with pressure altitude, true altitude, or absolute altitude. It is not to be used as a height reference, but as a determining criterion in the performance capability of an aircraft. Air density decreases with altitude. As air density decreases, density altitude increases. The further effects of high temperature and high humidity are cumulative, resulting in an increasing high density altitude condition. High density altitude reduces all aircraft performance parameters. To the pilot, this means that the normal horsepower output is reduced, propeller efficiency is reduced and a higher true airspeed is required to sustain the aircraft throughout its operating parameters. It means an increase in runway length requirements for takeoff and landings, and decreased rate of climb. An average small airplane, for example, requiring 1,000 feet for takeoff at sea level under standard atmospheric conditions will require a takeoff run of approximately 2,000 feet at an operational altitude of 5,000 feet.

departure control (DEP CON)—A function of approach control providing service for departing IFR aircraft and, on occasion, VFR aircraft.

departure procedure—Unless otherwise required, each pilot of an aircraft shall climb to 1,500 feet above surface as rapidly as practicable.

departure stall procedure—Nose down for airspeed, then proceed at proper climb attitude and speed.

depression—A low-pressure, or cyclonic, area.

descent rate—This varies with type of aircraft. A small aircraft may have a normal descent rate of 500 feet per minute, while a large commercial jet may approach an airport in a descent of 2,000 feet per minute.

deviation—The error produced in a magnetic compass b electrical equipment, and other factors in the aircraft.

deviations—

1. A departure from a current clearance, such as an off maneuver to avoid weather or turbulence.

2. Where specifically authorized in the FARs and requeste pilot, ATC may permit pilots to deviate from certain regulatio

dew—Atmospheric moisture condensed as liquid upon thos which are cooler than the surrounding air.

dew point-The temperature to which air must be coole stant pressure and moisture content in order for saturation Fog forms when the temperature-dew point spread is 5 Fahrenheit or less and decreasing.

DEWIZ—(*See* DISTANT EARLY WARNING IDENTIFICATION ZONE.)

DF—Direction finder equipment available at an airport.

DF GUIDANCE/DF STEER—Headings provided to aircraf ties equipped with direction finding equipment. These h followed, will lead the aircraft to a predetermined point s DF station or an airport. DF guidance is given to aircraft in to other aircraft which request the service. Practice DF g provided when workload permits. (*See* DIRECTION FINDER, D

DF fix—The geographical location of an aircraft obtained more direction finders.

direct tail wind on run-up area—Be sure to turn crosswind.

direction finder/DF/UDF/VDF/UVDF—A radio receiv with a directional sensing antenna used to take bearings transmitter. Specialized radio direction finders are used i air navigation aids. Others are ground based primarily to o on a pilot requesting orientation assistance or to locate craft. A location "fix" is established by the intersection of bearing lines plotted on a navigational chart using eithe ately located Direction Finders to obtain a fix on an ai pilot plotting the bearing indications of his DF on tw

directional finder/DF/UDF/VDF/UVDF ***continued***

located ground based transmitters both of which can be identified on his chart. UDFs receive signals in the ultra high frequency radio broadcast band; VDFs in the very high frequency band; and UVDFs in both bands. ATC provides DF service at those air traffic control towers and flight service stations listed in *Airport/Facility Directory* and DOD FLIP IFR En Route Supplement. (*See* DF GUIDANCE, DF FIX.)

direction, traffic pattern—(*See* TRAFFIC PATTERN DIRECTION.)

disaster area—That airspace below 2,000 feet above the surface within 5 miles of an aircraft or train accident, flood, forest fire, earthquake, and so on. Pilots require special permission to enter.

discrete code/discrete beacon code—As used in the Air Traffic Control Radar Beacon System (ATCRBS), any one of the 4096 selectable Mode 3/A aircraft transponder codes except those ending in zero zero; e.g., discrete codes: 0010, 1201, 2317, 7777; nondiscrete codes: 0100, 1200, 7700. Nondiscrete codes are normally reserved for radar facilities that are not equipped with discrete decoding capability and for other purposes such as emergencies (7700), VFR aircraft (1200), etc. (*See* RADAR.)

discrete frequency—A separate radio frequency for use in direct pilot-controller communications in air traffic control which reduces frequency congestion by controlling the number of aircraft operating on a particular frequency at one time. Discrete frequencies are normally designated for each control sector in en route/terminal ATC facilities. Discrete frequencies are listed in the Airport/Facility Directory, and DOD FLIP IFR En Route Supplement. (*See* CONTROL SECTOR.)

discontinuity—An atmospheric zone with relatively rapid transition of meteorological elements.

displaced threshold—A threshold that is located at a point on the runway other than the designated beginning of the runway. (*See* THRESHOLD.)

distance from clouds—(*See* VFR FLIGHT WEATHER MINIMUMS IN CONTROLLED AIRSPACE; VFR FLIGHT WEATHER MINIMUMS IN UNCONTROLLED AIRSPACE.)

distance measuring equipment (DME)—Airborne or ground equipment—UHF standard (TACAN compatible)—used to measure, in nautical miles, the slant range distance of an aircraft from a navigational aid (NAVAID).

distant early warning identification zone (DEWIZ)—An identification zone of defined dimensions extending upwards from the surface, in the DEW Line in Canada and around the entire coastal area of Alaska.

distress—An aircraft in distress has right-of-way over all other air traffic.

ditching over water—Use normal landing attitude, flaps down, slow speed (but keep power on, if available, for more control), prop door open with a map or other object, so it won't jam.

diurnal—Actions which recur daily, every 24 hours.

dive—A steep descent, with or without power, at a greater airspeed than for normal flight.

DME—(*See* DISTANCE MEASURING EQUIPMENT.)

DME fix—A geographical position determined by reference to a navigational aid. It provides distance in nautical miles and a radial in degrees magnetic from that NAVAID.

DME separation—Spacing aircraft in terms of distance determined by reference to DME.

Dmsh—Diminish.

Dnse—Dense.

documents required in an aircraft—Airworthiness certificate, registration certificate, operating manual, limitations placard, engine log, aircraft log, radio station license.

DOD FLIP—Department of Defense Flight Information Publications used for flight planning, en route, and terminal operations. FLIP is produced by the Defense Mapping Agency for worldwide use. United States Government Flight Information Publications (en route charts and instrument approach procedure charts) are incorporated in DOD FLIP for use in the National Airspace System (NAS).

doldrums—The equatorial belt of calm air or light variable winds lying between the two trade wind belts.

downwash—The downward thrust imparted on the air to provide lift for the airplane.

downwind leg—(*See* TRAFFIC PATTERN.)

drag—The resistance to flight built up by air particles passing around and over the aircraft.

drift—Deflection of an airplane from its intended course by the action of the wind.

drzl—Drizzle.

drugs—No person may act as a crewmember while using any drug which may affect his faculties in an unsafe way. Certain specific drugs which have been associated with aircraft accidents in the recent past are: antihistamines (widely prescribed for hay fever and other allergies); tranquillizers (prescribed for hypertension and other nervous conditions); reducing drugs (amphetamines and other appetite-suppressing drugs can produce sensations of wellbeing which have an adverse effect on judgment); barbiturates, nerve tonics or pills, prescribed for digestive and other disorders (barbiturates produce a marked suppression of mental alertness).

dry air—Denser than moist air, which has only ⅝ as much weight.

dsipt—Dissipate.

dsplcd—Displaced.

durn—Duration.

DVFR—Defense visual flight rule.

dvlp—Develop.

E

E in the phonetic alphabet is Echo (eck-oh).

E—Symbol for equatorial air mass.

ear, middle, discomfort or pain—Certain persons (whether pilots or passengers) have difficulty balancing the air loads on the eardrum while descending. This is particularly troublesome if a head cold or throat inflammation keeps the Eustachian tube from opening properly. If this trouble occurs during descent, try swallowing, yawning, or holding the nose and mouth shut and forcibly exhaling. If no relief occurs, climb back up a few thousand feet to relieve the pressure on the outer drum. Then descend again, using these measures. A more gradual descent may be tried, and it may be necessary to go through several climbs and descents to "stair step" down. If a nasal inhaler is available, it may afford relief. If trouble persists several hours after landing, consult an aviation medical examiner. NOTE: If you find yourself airborne with a head cold, you may avoid trouble by keeping an inhaler as part of the flight kit.

EAS—(*See* AIRSPEED.)

eddy—A whirling or circling current of air or water moving against the general flow.

efctv—Effective.

elev—Elevation.

elevation—Is given in feet above mean sea level and is based on highest usable portion of the landing area. When sea level, elevation will be indicated as "00"; when below, a minus sign (-) will precede the figure.

elevator—A hinged, horizontal control surface used to raise or lower the tail in flight.

eligibility requirements—(*See* PRIVATE PILOT, PREREQUISITES FOR CERTIFICATE; STUDENT PILOT CERTIFICATE.)

embdd—Embedded.

emerg—Emergency.

emergency locator transmitter (ELT)—A radio transmitter attached to the aircraft structure, operating from its own power source on 121.5 MHz and 243 MHz, transmitting a distinctively downward-swept audio tone for homing purposes, and designed to function without human action after an accident.

emergency procedures, general—When a pilot is in doubt of his position, or feels apprehensive for his safety, he should not hesitate to request assistance. Search and rescue facilities including radar, radio, and DF stations are ready and willing to help. There is no penalty for using them. Delay has caused accidents and cost lives. Safety is not a luxury. Take action!

If your radio is operative, contact any nearby radio facility or the emergency frequency 121.5 MHz, identifying your aircraft, heading and altitude, and your problem.

1. If equipped with a radar beacon transponder, and if unable to establish voice communications with an air traffic control facility, switch to Mode A/3 and Code 7700.

2. Comply with information and clearance received. Accept the communications control offered to you by the ground radio station, silence interfering radio stations, and do not shift frequency or shift to another ground station unless absolutely necessary.

3. If crash is imminent and equipped with a locator beacon, actuate the emergency signal.

emergency visual codes—[*See* Figure E-1(a) and (b).]

empennage—Term used to designate the entire tail group of an airplane, including both the fixed and movable tail surfaces.

eng—Engine.

engine compartment—Has an access door for preflight inspection for loose wires, clamps, oil, and fuel leaks.

GROUND-AIR VISUAL CODE FOR USE BY SURVIVORS

NO.	MESSAGE	CODE SYMBOL
1	Require assistance	V
2	Require medical assistance	X
3	No or Negative	N
4	Yes or Affirmative	Y
5	Proceeding in this direction	↑

IF IN DOUBT, USE INTERNATIONAL SYMBOL **SOS**

INSTRUCTIONS

1. Lay out symbols by using strips of fabric or parachutes, pieces of wood, stones, or any available material.
2. Provide as much color contrast as possible between material used for symbols and background against which symbols are exposed.
3. Symbols should be at least 10 feet high or larger. Care should be taken to lay out symbols exactly as shown.
4. In addition to using symbols, every effort is to be made to attract attention by means of radio, flares, smoke, or other available means.
5. On snow covered ground, signals can be made by dragging, shoveling or tramping. Depressed areas forming symbols will appear black from the air.
6. Pilot should acknowledge message by rocking wings from side to side.

Figure E-1(a). Emergency visual code—Survivors.

GROUND-AIR VISUAL CODE FOR USE BY GROUND SEARCH PARTIES		
NO.	MESSAGE	CODE SYMBOL
1	Operation completed.	LLL
2	We have found all personnel.	 LL (underlined)
3	We have found only some personnel.	 ++
4	We are not able to continue. Returning to base.	XX
5	Have divided into two groups. Each proceeding in direction indicated.	 [symbol: line with two arrows]
6	Information received that aircraft is in this direction.	– →
7	Nothing found. Will continue search.	NN

"Note: These visual signals have been accepted for international use and appear in Annex 12 to the Convention on International Civil Aviation."

Figure E-1(b). Emergency visual codes—Search parties.

engines—Usually in light aircraft of the opposed type—where pistons are arranged on opposite sides of the crankshaft.

engine-out emergency procedure—On departure lift off, land straight ahead. In the air, put aircraft in proper glide attitude, then if engine cannot be restarted: (1) by pilot, making sure ignition switch is set to both systems; (2) fuel selector is on a full tank; (3) by priming the engine; (4) making sure mixture is set to rich; and (5) carburetor heat is on; then he should pick best available landing spot, watching carefully for wind direction and overhead wires, and turn *off* all switches *just prior* to landing.

en route air traffic control service—Air traffic control service provided aircraft on an IFR flight plan between departure and destination terminal areas. When equipment capabilities and controller workload permit, certain advisory/assistance services may be provided to VFR aircraft.

en route automated radar tracking system (EARTS)—An automated radar and radar beacon tracking system. Its functional capabilites and design are essentially the same as the terminal ARTS IIIA system except for the EARTS capability of employing both short-range (ASR) and long-range (ARSR) radars, use of full digital radar displays, and fail-safe design. (*See* AUTOMATED RADAR TERMINAL SYSTEMS [ARTS].)

en route charts—(*See* AERONAUTICAL CHARTS.)

en route descent—Descent from the en route cruising altitude which takes place along the route of flight.

en route flight advisory service (EFAS)—

1. EFAS is a service specifically designed to provide en route aircraft with timely and meaningful weather advisories pertinent to the type of flight intended, route of flight and altitude. It is normally available throughout the conterminous U.S. from 6 A.M. to 10 P.M. at a service criterion of 5,000 feet above ground level. EFAS is provided by specially trained specialists from selected FSSs controlling one or more remote communications outlets covering a large geographical area. All communications are conducted on the designated EFAS frequency, 122.0 MHz.

en route flight advisory service (EFAS) ***continued***

2. To contact a flight watch facility on 122.0 MHz use the name of the VOR nearest your location and the words FLIGHT WATCH or, if the controlling facility is unknown, simply call "FLIGHT WATCH" and give the aircraft position.
Example: OAKLAND FLIGHT WATCH, LEAR TWO THREE FOUR FIVE KILO OVER.
Example: FLIGHT WATCH, COMMANDER FIVE SIX SEVEN LIMA FOXTROT OVER.

equip—Equipment.

equipment required—(*See* VISUAL FLIGHT RULES, REQUIRED EQUIPMENT, DAY; VISUAL FLIGHT RULES, REQUIRED EQUIPMENT, NIGHT.)

equivalent airspeed—(*See* AIRSPEED.)

ETA—Estimated time of arrival at a destination.

ETOV—Estimated time over a flight watch point, for a VFR pilot who has requested flight following service.

evaporation—The transformation of a liquid to the gaseous state. (The liquid loses heat during this process.)

examination for private pilot rating—Requires these prerequisites: pass a written test within 24 months before the flight test, pass a flight test, have the required flight experience (*See* PRIVATE PILOT, PREREQUISITES FOR CERTIFICATE) and have a third-class medical certificate.

expect (altitude) at (time) or (fix)—Used under certain conditions in a departure clearance to provide a pilot with an altitude to be used in the event of two-way communication failure.

expect further clearance (time)/EFC—The time a pilot can expect to receive clearance beyond a clearance limit.

expect further clearance via (airways, routes or fixes)—Used to inform pilot of the routing he can expect if any part of the route beyond a short range clearance limit differs from that filed.

expected approach clearance time (EAC)—The time at which it is expected that an arriving aircraft will be cleared to begin approach for a landing.

expected further clearance time (EFC)—The time at which it is expected that additional clearance will be issued to an aircraft.

expedite—Used by ATC when prompt compliance is required to avoid the development of an imminent situation.

experience requirements—(*See* SOLO FLIGHT REQUIREMENT; NIGHT FLIGHT EXPERIENCE; RECENT FLIGHT EXPERIENCE; PRIVATE PILOT, PREREQUISITES FOR CERTIFICATE.)

extd—Extend, extended.

extrm—extreme.

extsv—extensive.

F

F in the phonetic alphabet is Foxtrot (foks-trot).

FA—(*See* AREA FORECAST.)

Fahrenheit—Temperature scale on which, at standard atmospheric pressure, 32 degrees denotes the freezing temperature and 212 degrees the temperature of boiling water.

fall wind—A wind blowing down a mountainside (also called gravity wind).

false cirrus—(*See* CLOUDS.)

FAR—Federal Aviation Regulations.

fatigue—Generally slows reaction times and causes inattention, resulting in foolish errors. In addition to the most common cause of fatigue, insufficient rest and loss of sleep, the pressures of business, financial worries, and family problems can be important contributing factors. If fatigue is marked prior to a given flight, don't fly. To prevent fatigue effects during long flights, keep active with respect to making ground checks, radio-navigation position plotting, and remaining mentally active.

FBO—Fixed base operator at an airport, operating services such as fueling, tiedown, maintenance, a restaurant, and so on.

fcst—Forecast.

feather the prop—Adjust the pitch of the blades of a propeller whose power has been cut off, so they will offer least resistance to oncoming air.

federal airways, use of—Pilots not operating on an IFR flight plan, and when in level cruising flight, are cautioned to conform with VFR cruising altitudes appropriate to direction of flight. During climb or descent, pilots are encouraged to fly to the right side of the center line of the radial forming the airway in order to avoid IFR and VFR cruising traffic operating along the center line of the airway.

Federal Aviation Administration—Establishes regulations for the safe operation of private, business, commercial, and military aircraft in the United States. Washington address is Federal Aviation Administration, Department of Transportation, Washington, DC 20591.

figures—(phraseology in communication) Indicates hundreds and thousands in round number, as for ceiling heights and upper wind levels.

1. Figures up to 9900 shall be spoken in accordance with the following examples: 500—FIVE HUNDRED; 1300—ONE THOUSAND THREE HUNDRED; 4500—FOUR THOUSAND FIVE HUNDRED; 9900—NINER THOUSAND NINER HUNDRED.

2. Numbers above 9900 shall be spoken by separating the digits preceding the word "thousand," as follows: 10000—ONE ZERO THOUSAND; 13000—ONE THREE THOUSAND; 18500—ONE EIGHT THOUSAND FIVE HUNDRED; 27000—TWO SEVEN THOUSAND.

3. Transmit airway or jet route numbers as follows: V12—VICTOR TWELVE; J533—J FIVE THIRTY THREE.

4. All other numbers shall be transmitted by pronoucing each digit. Examples: 10—ONE ZERO; 75—SEVEN FIVE; 583—FIVE EIGHT THREE; 1850—ONE EIGHT FIVE ZERO; 18143—ONE EIGHT ONE FOUR THREE; 26075—TWO SIX ZERO SEVEN FIVE. The digit "9" shall be spoken "NINER."

5. When a radio frequency contains a decimal point, the decimal point is spoken as "POINT." Examples: 122.1—ONE TWO TWO

POINT ONE; 126.7—ONE TWO SIX POINT SEVEN. (International Civil Aviation Organization procedures require the decimal point be spoken as "DECIMAL," and Federal Aviation Administration will honor such usage by military aircraft and all other aircraft required to use ICAO procedures.)

final approach—Aircraft on final approach or landing has right of way over other aircraft in flight or on surface. When two aircraft are approaching airport, the craft at lower altitude has right of way.

final approach (VFR)—A flight path of a landing aircraft in the direction of landing along the extended runway centerline from the base leg to the runway.

final controller—That controller providing final approach guidance utilizing radar equipment.

first class medical certificate—(*See* MEDICAL CERTIFICATES.)

fix—A geographical position determined by visual reference to the surface, by reference to one or more radio navigational aids, by celestial plotting, or by another navigational device.

fixed pitch—Type of aircraft propeller whose action is constant under all flying conditions and cannot be changed.

FL—Flight level.

flag/flag alarm—A warning device incorporated in certain airborne navigation and flight instruments indicating that:

1. Instruments are inoperative or otherwise not operating satisfactorily, or
2. Signal strength or quality of the received signal falls below acceptable values.

flameout—Unintended loss of combustion in turbine engines resulting in the loss of engine power.

flammable—Susceptible to igniting readily or exploding.

flap extended speed—Highest speed permissible with wing flaps in prescribed extended position.

flaps—Change the lifting capacity of the wing by changing the area of the wing surface. An increased lifting capacity (flaps down) permits

flaps *continued*

slower landing and takeoff speed. *After* sufficient speed and altitude are gained, retracting flaps then reduces drag at high speed.

flare out—To round out a landing by decreasing the rate of descent and air speed by slowly raising the nose (pulling back slowly on the elevator).

flat out—Flying at full throttle.

fld—Field.

flight assists—In 1982, flight service stations, terminals, and centers handled 959 flight assists, the providing of maximum assistance to an aircraft in distress. With very few exceptions, a flight assist results in successful saving of aircraft. The statistics on flight assists should not be confused with the data on aircraft accidents. Figures for 1980 show that general aviation pilots were involved in 3,799 accidents, resulting in 1,375 fatalities.

Why were these 959 pilots involved in flight assist situations? The "types of situations" are listed below. More than one type is possible per pilot: lost—560; weather related—210; fuel—231; equipment failure—108; gear up—72.

A look at the first three factors above (lost, weather, fuel) indicates improper preflight planning and lack of knowledge and weather briefing. DON'T BECOME A STATISTIC. Check the weather, the rules, your charts; file a flight plan. And last but not least, check that aircraft!

flight altitudes—(phraseology in communications):

1. Up to but not including 18,000′ MSL—the separate digits of the thousands, plus the hundreds, if appropriate. Examples: 12,000—ONE TWO THOUSAND; 12,500—ONE TWO THOUSAND FIVE HUNDRED.

2. At and above 18,000′ MSL (FL 180) the words "flight level" followed by the separate digits of the flight level. Examples: FL 190—FLIGHT LEVEL ONE NINER ZERO; FL 275—FLIGHT LEVEL TWO SEVEN FIVE.

"flight check" aircraft in terminal areas—

1. Flight Check is a call sign used to alert pilots and air traffic controllers when a FAA aircraft is engaged in flight inspection/certification of navaids and flight procedures. Flight Check aircraft fly pre-

planned high/low altitude flight patterns such as grids, orbits, DME arcs, and tracks, including low passes along the full length of the runway to verify navaid performance. In most instances, these flight checks are being automatically recorded and/or flown in an automated mode.

2. Pilots should be especially watchful and avoid the flight paths of any aircraft using the call sign "Flight Check" or "Flight Check recorded." The latter call sign, e.g., "Flight Check 47 recorded" indicates that automated flight inspections are in progress in terminal areas. These flights will normally receive special handling from ATC. Pilot patience and cooperation in allowing uninterrupted recordings can significantly help expedite flight inspections, minimize costly, repetitive runs, and reduce the burden on the U.S. taxpayer.

flight experience—(*See* PREREQUISITES FOR PRIVATE PILOT CERTIFICATE; also RECENT FLIGHT EXPERIENCE; and NIGHT FLIGHT EXPERIENCE.)

flight instruments—(basic) These include: airspeed indicator, altimeter, compass, turn and slip indicator, horizon indicator, and rate of climb/descent indicator.

flight level (FL)—A level of constant atmospheric pressure related to a reference datum of 29.92 inches of mercury. Each is stated in three digits representing hundreds of feet; for example, FL 250 represents a barometric altimeter indication of 25,000 feet, FL 255 indicates 25,500 feet.

flight maneuvers—As indicated on cockpit instruments. (*See* Figure F-1.)

flight outside the United States—A flight plan is usually required to be filed by the pilot to the countries visited or overflown. Also, United States customs regulations must be met on return. For further information write for a copy of *Customs Guide for Private Flyers* to U.S. Customs Service, 1301 Constitution Ave., N.W., Washington, DC 20229, Attention: Public Information Officer.

flight plan—Specified information, filed orally or in writing with an air traffic control facility, relating to the intended flight of an aircraft. While a flight plan is not required for VFR flights unless the flight is to penetrate an ADIZ or DEWIZ, it is recommended as good operating practice. (*See* Figure F-2.)

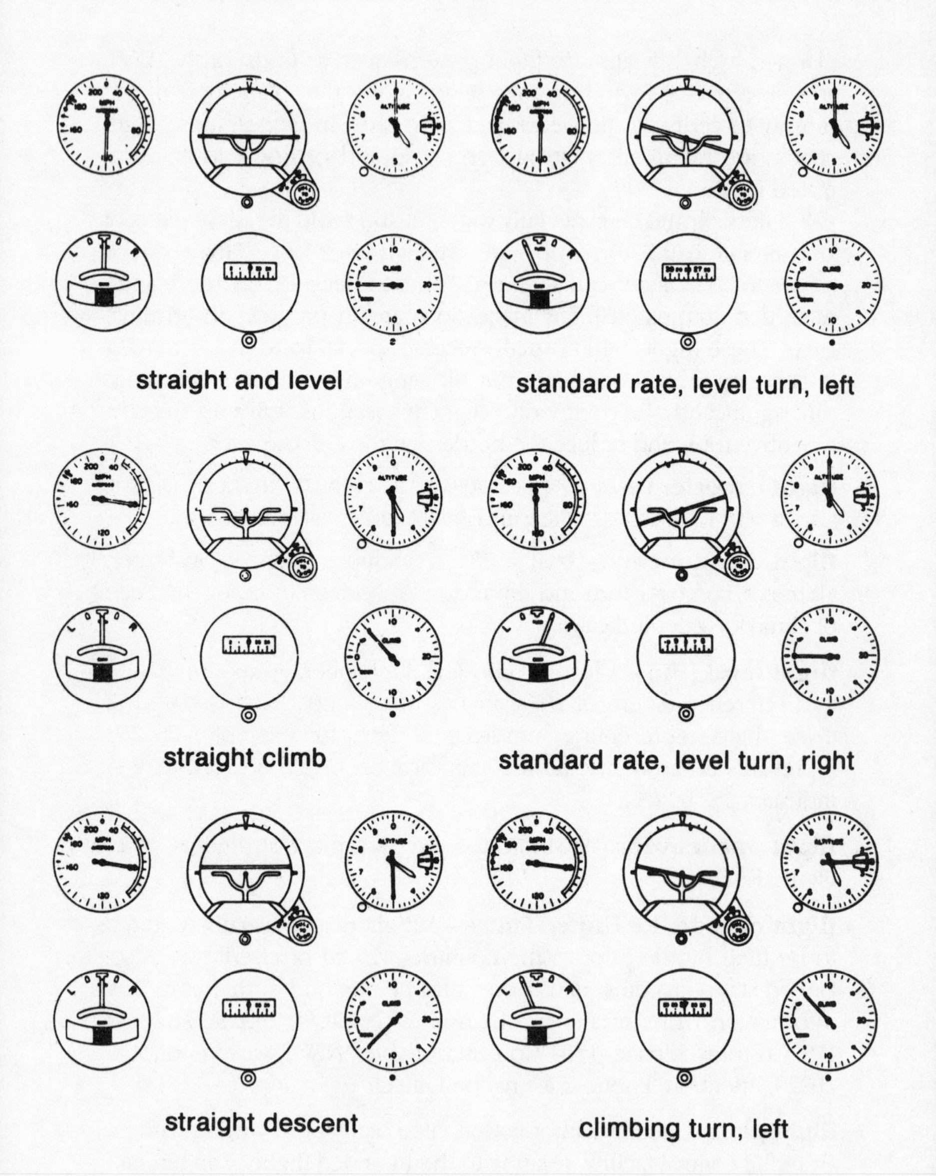

Figure F-1. Cockpit instruments indicating flight maneuvers.

Form Approved: OMB No. 04-R0072

DEPARTMENT OF TRANSPORTATION
FEDERAL AVIATION ADMINISTRATION

FLIGHT PLAN

CIVIL AIRCRAFT PILOTS. FAR Part 91 requires you file an IFR flight plan to operate under instrument flight rules in controlled airspace. Failure to file could result in a civil penalty not to exceed $1,000 for each violation (Section 901 of the Federal Aviation Act of 1958, as amended). Filing of a VFR flight plan is recommended as a good operating practice. See also Part 99 for requirements concerning DVFR flight plans.

1. TYPE	2. AIRCRAFT IDENTIFICATION	3. AIRCRAFT TYPE/ SPECIAL EQUIPMENT	4. TRUE AIRSPEED	5. DEPARTURE POINT	6. DEPARTURE TIME		7. CRUISING ALTITUDE
					PROPOSED (Z)	ACTUAL (Z)	
VFR							
IFR							
DVFR			KTS				

8. ROUTE OF FLIGHT

9. DESTINATION (Name of airport and city)	10. EST. TIME ENROUTE		11. REMARKS
	HOURS	MINUTES	

12. FUEL ON BOARD		13. ALTERNATE AIRPORT(S)	14. PILOT'S NAME, ADDRESS & TELEPHONE NUMBER & AIRCRAFT HOME BASE	15. NUMBER ABOARD
HOURS	MINUTES			

16. COLOR OF AIRCRAFT

CLOSE VFR FLIGHT PLAN WITH ________________ FSS ON ARRIVAL

FAA Form 7233-1 (5-77)

Figure F-3. Flight plan form.

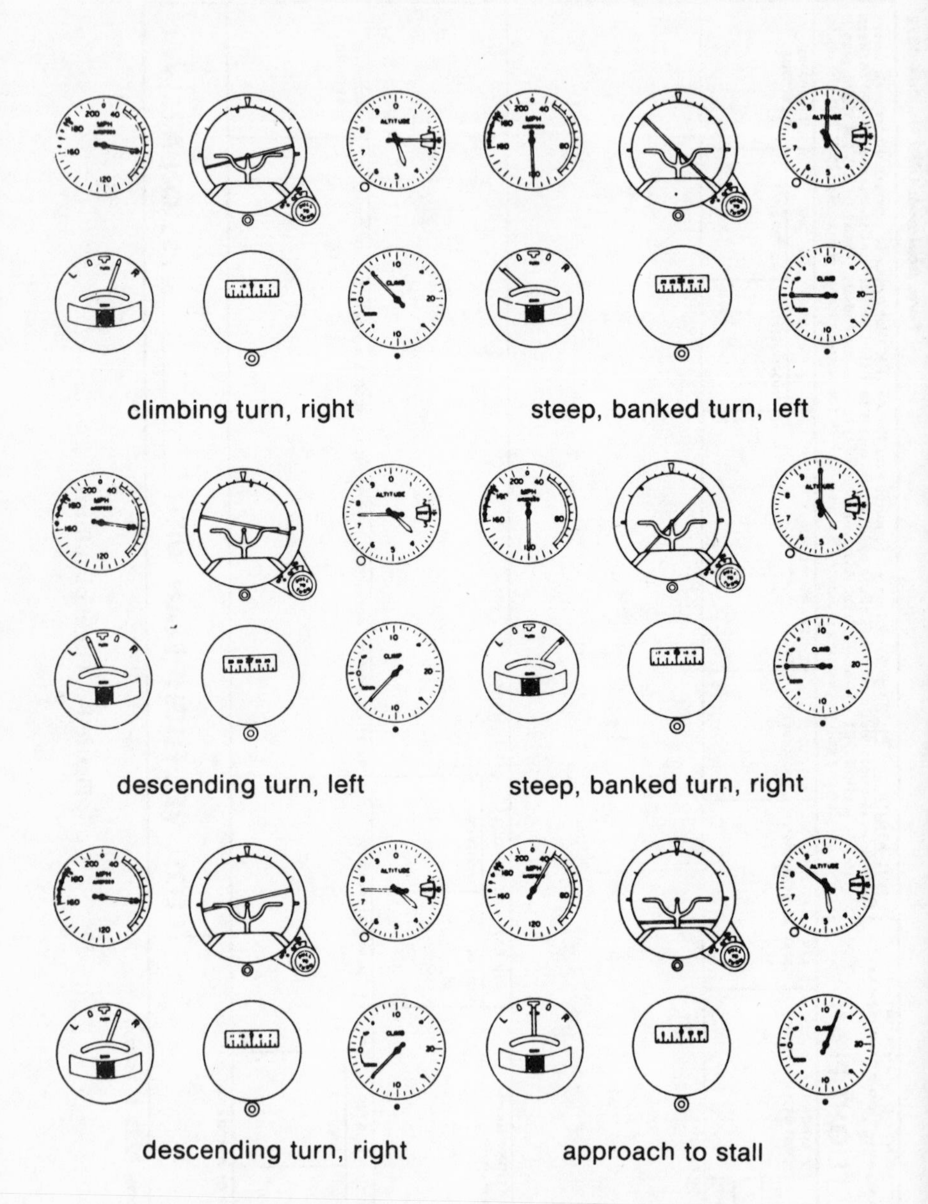

Figure F-2. Cockpit instruments indicating flight maneuvers.

flight plan, change in—In addition to altitude/flight level, destination and/or route changes, increasing or decreasing the speed of an aircraft constitutes a change in a flight plan. Therefore, if at any time the average true airspeed at cruising altitude between reporting points varies, or is expected to vary from that given in the flight plan by plus or minus 5 percent or 10 knots whichever is greater, air traffic control should be advised.

flight plans, closing VFR/DVFR—A pilot is responsible for ensuring that his VFR or DVFR flight plan is canceled (*See* FAR—Part 91.). You should close your flight plan with the nearest flight service station, or if one is not available you may request any ATC facility to relay your cancelation to the FSS. Control towers do not automatically close VFR or DVFR flight plans as they may not be aware that a particular VFR aircraft is on a flight plan. If you fail to report or cancel your flight plan within half an hour after your ETA, search and rescue procedures are started. (*See* VFR FLIGHT PLAN.)

flight proficiency—(*See* VFR FLIGHT PROFICIENCY.)

flight service stations (FSS)—Facilities operated by the Federal Aviation Administration to provide flight assistance service. They are within the National Airspace System and have the prime responsibility for preflight pilot briefing, relaying ATC clearances, en route communications with VFR flights, assisting lost VFR aircraft, originating NOTAMs, broadcasting aviation weather information, accepting and closing flight plans, monitoring radio NAVAIDs, participating with search and rescue units in locating missing VFR aircraft. In addition, at selected locations, FSSs take weather observations, issue airport advisories, provide en route flight advisory service (flight watch), and advise Customs and Immigration of transborder flights.

flight tests, prerequisites for—Applicant must have passed written test within 24 months prior to date he takes flight test and have applicable medical certificate and prescribed aeronautical experience. (*See* PRIVATE PILOT, PREREQUISITES FOR CERTIFICATE.)

flight time, logging of—Flight time used to meet experience (or recent flight) requirements must be shown by a reliable record (verified by an instructor or local or distant airport operator).

flight visibility—Average forward horizontal distance from cockpit of aircraft in flight at which prominent objects can be seen in day, and prominent lighted objects be seen at night.

flight watch—A shortened term for use in air-ground contacts on frequency 122.0 MHz to identify the flight service station providing En Route Flight Advisory Service; e.g., "Oakland Flight Watch." (*See* EN ROUTE FLIGHT ADVISORY SERVICE.)

FLIP—Flight information publication. (*See* DOD FLIP.)

flry—Flurry.

fly heading (degrees)—Controller informs the pilot of the heading he should fly. The pilot may have to turn to, or continue on, a specific compass direction in order to comply with the instructions. The pilot is expected to turn in the shorter direction to the heading, unless otherwise instructed by ATC.

flying clubs—Members who cooperatively own aircraft can sometimes offer the pilot a lower hourly rental rate than a commercial firm.

flying the beam—Flying along a radio beam used in aerial navigation.

flying the needles—Instrument flight.

FM—Fan marker.

FM radio receivers, portable—May not be operated in any commercial aircraft or any aircraft equipped with VOR equipment while it is being used for navigational purposes.

foehn—Unseasonably warm, dry wind with a strong downward component, found in many mountainous regions. (In the Rocky Mountains, called a chinook.)

fog—A cloud of water droplets at or near the earth's surface.

fone—Telephone.

forced landing—Rule number one is never turn your back on a field on which you want to land. If you are too high, make S-turns, always turning back again toward the field until glide angle is right to land.

fornn—Forenoon.

Table F-1

TRANSMITTING AND RECEIVING FREQUENCIES

Frequency	Use
118.0-121.4	Air Traffic Control
121.425-121.475	Band Protection for 121.5
121.5	Emergency Search and Rescue (ELT Operational Check, 5 Sec)
121.525-121.575	Band Protection for 121.5
121.6-121.925	Airport Utility and ELT Test
121.95	Aviation Instructional
121.975	Private Aircraft Advisory (FSS)
122.0	En Route Flight Advisory Service (EFAS)
122.025-122.075	FSS
122.1	FSS Usually Receive Only Associated with VOR (May Be Simplex)
122.125-122.175	FSS
122.2	FSS Common En Route Simplex
122.225-122.675	FSS
122.7	UNICOM—Uncontrolled Airports
122.725	UNICOM
122.75	UNICOM—Private Airports (Not Open to the Public) and Air-to-Air
122.775	UNICOM
122.8	UNICOM
122.825	UNICOM
122.85	Multicom—Special Use
122.875	UNICOM
122.9	Multicom—Special Use
122.925	Multicom—Natural Resources
122.950	UNICOM—Airports with a Control Tower
122.975	UNICOM—High Altitude Above 10,000 feet (3000 m)
123.0	UNICOM—Uncontrolled Airports
123.025	Helicopter Air-to-Air
123.05	UNICOM—Heliports
123.075	UNICOM—Heliports
123.1	Search and Rescue (Temporary Control Towers, Fly-ins may be assigned on a non interference basis to Search and Rescue.)
123.125-123.275	Flight Test
123.3	Aviation Instructional—Gliders
123.325-123.475	Flight Test
123.5	Aviation Instructional—Gliders
123.525-123.575	Flight Test
123.6	FSS
123.625	Air Traffic Control
123.65	FSS
123.675-126.175	Air Traffic Control
126.2	Air Traffic Control—Military (Common)
126.225-128.8	Air Traffic Control
128.825-132.0	Operational Control (ARINC)
132.025-134.075	Air Traffic Control
134.1	Air Traffic Control—Military (Common)
134.125-135.825	Air Traffic Control
135.85	Flight Inspection
135.875-135.925	Air Traffic Control
135.95	Flight Inspection
135.975	Air Traffic Control

four C's—(*See* C'S, THE FOUR.)

fqt—Frequent.

freq—Frequency.

frequency in use—(*See* Table F-1.)

frequency congestion—(*See* RADIO FREQUENCY CONGESTION.)

front—The zone of transition between two air masses of different density.

fropa—Frontal passage.

frosfc—Frontal surface.

frost—Crystals of ice formed like dew, but at a temperature below freezing.

FSS—Flight service station.

FSS, to contact an—Flight service stations are allocated frequencies for different functions. For airport advisory service the pilot should contact the FSS on the frequency listed in the *Airport Facility Directory.* If you are in doubt as to what frequency to use to contact an FSS, transmit on 122.2 MHz and advise them of the frequency you are receiving on.

fuel—Octane rating proper for type of aircraft must always be used. (*See* OWNER'S MANUAL.)

fuel contamination—Can be caused by (1) servicing the aircraft from improperly filtered tanks and (2) storing an aircraft with only partially filled fuel tanks (which can cause condensation and water contamination).

fuel requirements for flight under VFR—No person may begin a flight in an airplane under VFR unless (considering wind and forecast weather conditions) there is enough fuel to fly to the first point of intended landing and, assuming normal cruising speed—

1. During the day, to fly after that for at least 30 minutes; or
2. At night, to fly after that for at least 45 minutes.

fuel siphoning/fuel venting—Unintentional release of fuel caused by overflow, puncture, loose cap, etc.

fuel symbols—

CODE	FUEL
80	Grade 80 gasoline (Red)
100	Grade 100 gasoline (Green)
100LL	Grade 100LL gasoline (low lead) (Blue)
115	Grade 115 gasoline
A	Jet A—Kerosene freeze point—40° C.
A1	Jet A-1—Kerosene freeze point—50° C.
A1+	Jet A-1—Kerosene with icing inhibitor, freeze point—50° C.
B	Jet B—Wide-cut turbine fuel, freeze point—50° C.
B+	Jet B—Wide-cut turbine fuel with icing inhibitor, freeze point—50° C.

fuse panels—Are located in most aircraft and should be checked if electrical equipment fails to function.

fuselage—The body of the aircraft to which the wings, landing gear, and tail are attached.

G

G in the phonetic alphabet is Golf, pronounced like the sport.

GCA—Ground controlled approach.

general aviation—That portion of civil aviation which encompasses all facets of aviation except air carriers holding a certificate of public convenience and necessity from the U.S. Department of Transportation, and large aircraft commercial operators.

giving way—(*See* ACCIDENT CAUSE FACTORS.)

glaze—A coating of ice, generally clear and smooth, which forms on exposed objects by the freezing of supercooled water deposited by rain, drizzle, and fog, and sometimes condensed from supercooled water vapor.

glide—Sustained forward flight in which power is off and speed is maintained only by the loss of altitude.

glide ratio—A glide ratio of 7.1 means the aircraft will glide a distance of 7 times its vertical height. That is, if it is 1 mile high, with power off under "no wind" conditions, it would glide 7 miles. Glide ratios vary depending on the type of aircraft.

glide slope (GS)—Provides vertical guidance for aircraft during approach and landing. The glide slope consists of the following:

1. Electronic components emitting signals which provide vertical guidance by reference to airborne instruments during instrument approaches such as an ILS, or

2. Visual ground aids, such as VASI, which provide vertical guidance for a VFR approach or for the visual portion of an instrument approach and landing.

gnd—Ground.

gndfg—Ground fog.

go ahead—Controller's expression meaning proceed with your message. Not to be used for any other purpose.

go around—Instructions for a pilot to abandon his approach to landing. Additional instructions may follow. Unless otherwise advised by ATC, a VFR aircraft or an aircraft conducting a visual approach should over-fly the runway while climbing to traffic pattern altitude and enter the traffic pattern via the crosswind leg. A pilot on an IFR flight plan making an instrument approach should execute the published missed approach procedure or proceed as instructed by ATC, e.g., "Go around" (additional instructions, if required).

grainy ice—(rime) looks like shaved ice. On the plane it increases the amount of weight to be carried and reduces the amount of lift quality of the wings.

graphic notices and supplemental data—A publication designed primarily as a pilot's operational manual containing a tabulation of parachute jump areas, special notice area graphics, terminal radar service area graphics, civil flight test areas, military refueling tracks and areas, and other data not requiring frequent change.

gravity—The attraction between the weight of an object and the earth.

gravity wind—A cold wind blowing downslope on a mountain.

grdl—Gradual.

green, flashing—Light signal from the tower meaning (on ground) cleared to taxi or (in flight) return for landing (to be followed by steady green at proper time).

green light—Alternating with red, general warning signal from tower; exercise extreme caution.

green, steady—Light signal from the tower meaning (in flight) either cleared for landing or (on ground) cleared for takeoff.

grid system—Degrees of latitude and longitude as marked off on maps to indicate geographic position on the earth.

ground-air visual code—For use by survivors and ground search parties. (*See* EMERGENCY VISUAL CODES.)

ground control radio communication phraseology—The following phraseologies and procedures are used in radio-telephone communications with aeronautical ground stations: aircraft identification, location, type of operation planned (VFR or IFR), and the point of first intended landing. Example: *Aircraft:* "WASHINGTON GROUND BEECHCRAFT ONE THREE ONE FIVE NINER AT HANGAR EIGHT, READY TO TAXI, VFR TO CHICAGO OVER." *Tower:* "BEECHCRAFT ONE THREE ONE FIVE NINER, RUNWAY THREE SIX, WIND THREE ZERO DEGREES AT TWO FIVE, ALTIMETER THREE ZERO ZERO FOUR, HOLD SHORT OF RUNWAY THREE."

ground control radio communications—Pilots of departing aircraft should communicate with the control tower, on the appropriate ground control frequency, for taxi and clearance information and, unless otherwise advised, should remain on that frequency until they are ready to request takeoff clearance. A pilot who has just landed should not change from the tower frequency to the ground control frequency until he is directed to do so by the controller.

Ground control frequencies are provided in the 121.6–121.9 MHz band. The controller may omit the frequency, or the numbers preceding the decimal point in the frequency, when directing the pilot to change to a VHF ground control frequency if, in the controller's opinion, this usage will be clearly understood by the pilot; e.g., 121.7—"Contact ground" or "Contact ground point seven."

ground effect—The temporary gain in lift during flight at very low altitudes due to the compression of air between the wings of the airplane and the ground.

ground fog—Also called radiation fog, is formed by the cooling of the land surfaces by radiation (loss of heat to the sky).

ground loop—An uncontrollable violent turnaround on the ground.

ground reference maneuvers—Over a path determined by points or lines on the ground, designed to help the student pilot, especially in maneuvering incident to landing approaches and departures. Basic maneuvers include S turns across a road, turns about a point, figure eights, and rectangular courses or traffic patterns.

groundspeed—The speed of the aircraft relative to the surface of the earth.

ground station call signs—Shall comprise the name of the location or airport, followed by the appropriate indication of the type of station: SHANNON UNICOM (airport unicom); OAKLAND TOWER (airport traffic control tower); MIAMI GROUND (ground control position in tower); DALLAS CLEARANCE DELIVERY (IFR clearance delivery position); KENNEDY APPROACH (tower radar or nonradar approach control position); ST. LOUIS DEPARTURE (tower radar departure control position); WASHINGTON RADIO (Federal Aviation Administration flight service station); SEATTLE FLIGHT WATCH (FAA Flight Service Station—En Route Flight Advisory Service on weather); NEW YORK CENTER (Federal Aviation Administration air route traffic control center).

ground visibility—Prevailing horizontal visibility near earth's surface as reported by weather bureau or an accredited observer.

guard—Listen to.

gust—A sudden, brief increase in the speed of the wind.

GWT—Gross weight.

gyro, directional—Unlike the compass, the magnetic heading does not show unless the compass card is set by a caging knob to agree with the compass. Then, since it has a tendency to creep off heading, it must be reset approximately every 15 minutes. In this way it is more stable for reading heading than the magnetic compass, as the latter is more seriously affected by flight maneuvers.

gyro horizon—Flight instrument showing the plane's altitude at a glance.

H

H in the phonetic alphabet is Hotel, pronounced (hoh-tel).

HAA—Height above airport.

hand signals in use at airports—(*See* Figure H-1.)

handoff—An action taken to transfer the radar identification of an aircraft from one controller to another if the aircraft will enter the receiving controller's airspace and radio communications with the aircraft will be transferred.

HAT—Height above touchdown.

have numbers—Used by pilots to inform ATC that they have received runway, wind, and altimeter information only.

hazardous area reporting service—Selected FSSs provide flight monitoring where regularly traveled VFR routes cross large bodies of water, swamps, and mountains. This service is provided for the purpose of expeditiously alerting Search and Rescue facilities when required.

1. When requesting the service either in person, by telephone, or by radio, pilots should be prepared to give the following information—type of aircraft, altitude, indicated airspeed, present position, route of flight, heading.

2. Radio contacts are desired at least every 10 minutes. If contact is lost for more than 15 minutes, Search and Rescue will be alerted. Pilots are responsible for canceling their request for service when they are outside the service area boundary. Pilots experiencing two-way radio failure are expected to land as soon as practicable and cancel their request for service. (*See* LAKE, ISLAND, MOUNTAIN, AND SWAMP REPORTING SERVICE, FIGURE L-1.)

Figure H-1. Hand signals in use at airports.

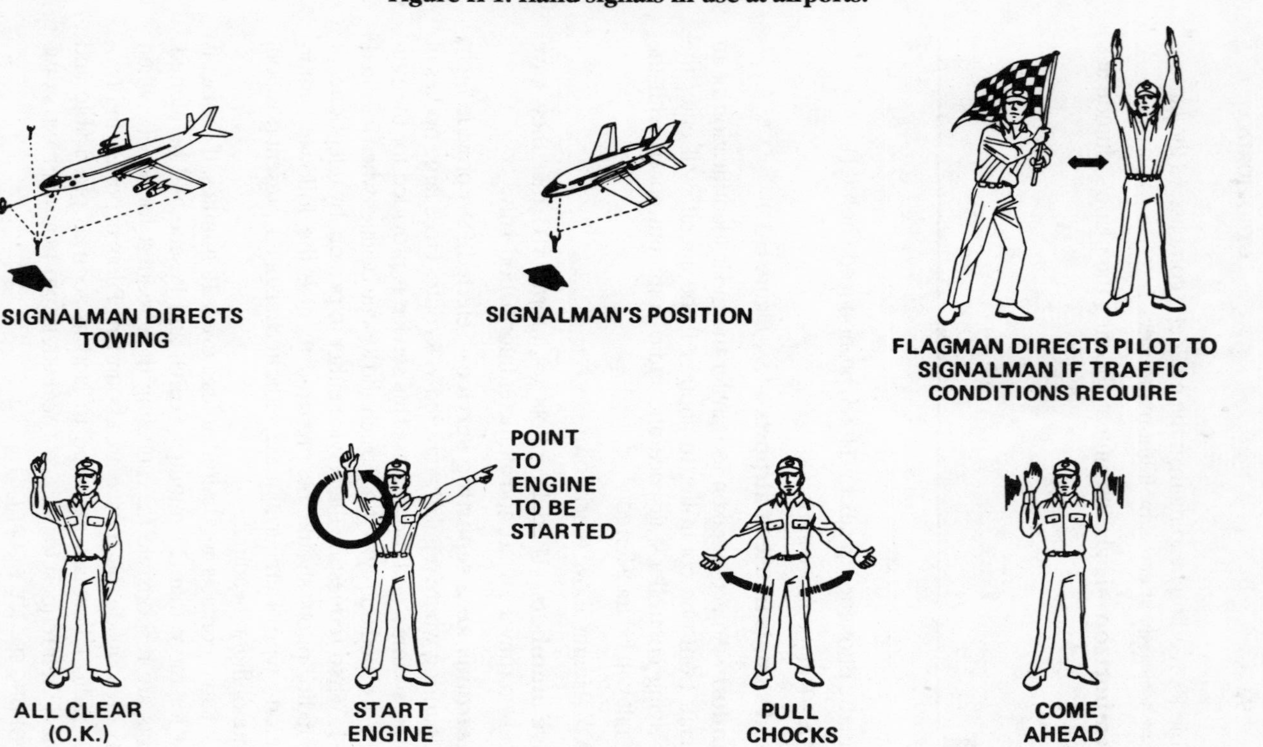

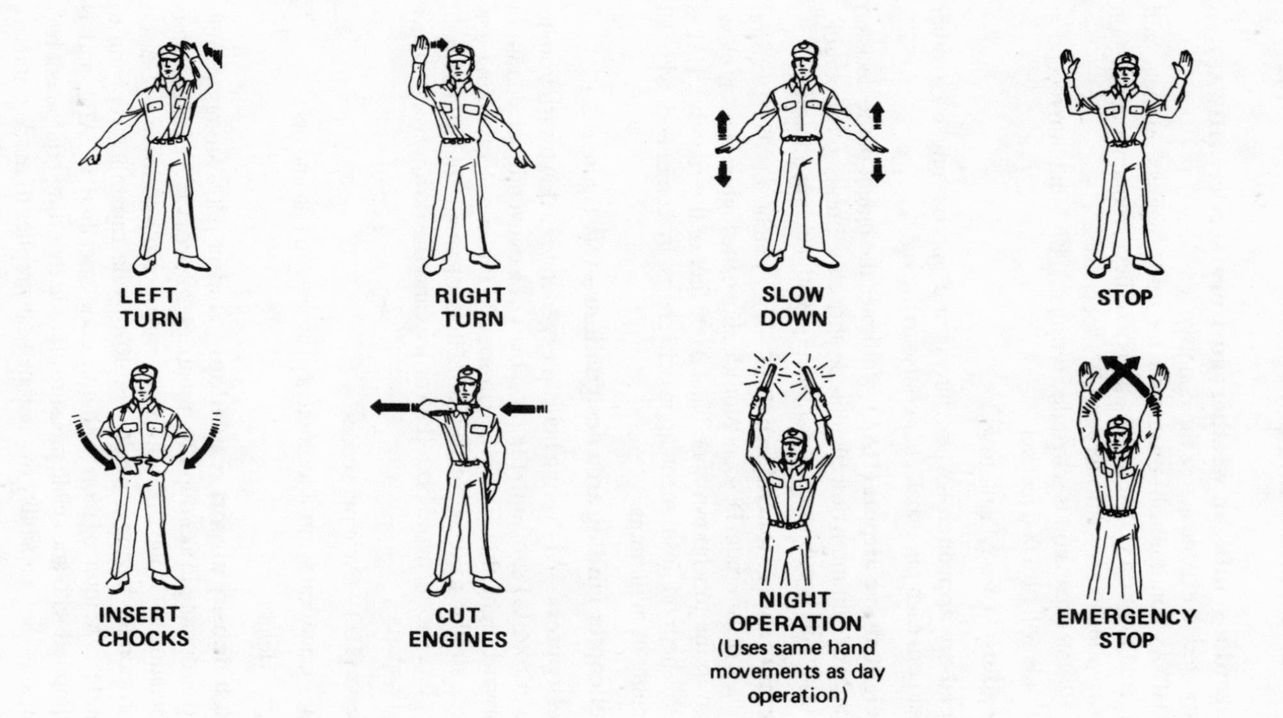
LEFT TURN
RIGHT TURN
SLOW DOWN
STOP
INSERT CHOCKS
CUT ENGINES
NIGHT OPERATION (Uses same hand movements as day operation)
EMERGENCY STOP

hazardous inflight weather advisory service (HIWAS)—is expected to expand across the country.

HIWAS continuously broadcasts recorded Sigmets, Airmets, and urgent Pireps on VOR voice channels to relieve frequency congestion and workload on controllers and FSS specialists.

HIWAS expansion is scheduled through 1987. Minimum reception altitude will be 4,000 feet agl.

heading—(*See* TRUE HEADING.)

head-on approach—When aircraft are approaching each other head-on, each pilot shall alter course to the right.

height above airport (HAA)—Indicates the height of the decision height or minimum descent altitude above the published airport elevation. This is published in conjunction with circling minimums.

height above touchdown/HAT—The height of the Decision Height or Minimum Descent Altitude above the highest runway elevation in the touchdown zone (first 3,000 feet of the runway). HAT is published on instrument approach charts in conjuntion with all straight-in minimums.

helicopter landing area designations—(*See* Figure H-2.)

helicopters—A hovering helicopter generates a downwash from its main rotor(s) similar to the prop blast of a conventional aircraft. In forward flight, this energy is transformed into a pair of trailing vortices similar to those of wingtip. Pilots of small aircraft and helicopters should avoid both the vortices and the downwash of heavy helicopters.

hertz (Hz)—Cycle per second.

Hg—Chemical symbol for mercury (as used in barometers).

hgt—Height.

high density airport regulations—Student pilots are not permitted to operate in a Group I Terminal Control Area (TCA). Private pilots operating in Group I TCAs must have an operable coded radar beacon transponder having a mode 3/A 4096 code capability, replying to mode 3/A interrogation with the code specified by ATC, and is equipped with automatic pressure altitude reporting equipment having a mode C capability that automatically replies to mode C interro-

Figure H-2. Helicopter landing areas.

gations by transmitting pressure altitude information in 100-foot increments. Private pilots must receive authorization for ATC prior to operating in the Terminal Control Area.

high frequency communications/HF communications—High radio frequencies (HF) between 3 and 30 MHz used for air-to-ground voice communication in overseas operations.

high performance aircraft in terminal areas—The Federal Aviation Administration has initiated a program known as "KEEP-'EM-HIGH." Arriving at terminal areas, high performance aircraft will be kept at the highest possible altitude for as long as possible. Departing high performance aircraft will be climbed to the highest possible altitude filed by the pilot as soon as possible. This program is intended to reduce the mixture of uncontrolled aircraft with high performance controlled aircraft in the vicinity of the airport.

high speed taxiway/exit/turnoff—A long radius taxiway designed and provided with lighting or marking to define the path of aircraft, traveling at high speed (up to 60 knots), from the runway center to a point on the center of a taxiway. [Also referred to as long radius exit or turn-off taxiway.] The high speed taxiway is designed to expedite aircraft turning off the runway after landing, thus reducing runway occupancy time.

hire—(*See* PASSENGERS, CARRYING OF.)

HIRL—High intensity runway lights.

hlsto—Hailstones.

hnd—Hundred.

hold/holding procedure—A predetermined maneuver which keeps aircraft within a specified airspace while awaiting further clearance from air traffic control. Also used during ground operations to keep aircraft within a specified area or at a specified point while awaiting further clearance from air traffic control. (*See* HOLDING FIX.)

holding fix—A specified fix identifiable to a pilot by NAVAIDS or visual reference to the ground used as a reference point in establishing and maintaining the position of an aircraft while holding.

homing—Flight toward a NAVAID, without correcting for wind, by adjusting the aircraft heading to maintain a relative bearing of zero degrees.

horsepower—A unit for measurement of power output of an engine. 1 h.p. = power required to raise 550 pounds 1 foot in 1 second.

how do you hear me?—A question relating to the quality of the transmission or to determine how well the transmission is being received.

hr—Hour.

hurcn—Hurricane.

hvy—Heavy.

hyperventilation—Or over breathing, is a disturbance of respiration that may occur in individuals as a result of emotional tension or anxiety. Under conditions of emotional stress, fright, or pain, breathing rate may increase, causing increased lung ventilation, although the carbon dioxide output of the body cells does not increase. As a result, carbon dioxide is "washed out" of the blood. The most common symptoms of hyperventilation are dizziness, hot and cold sensations, tingling of the hands, legs, and feet, tetany (spasms of the muscles), nausea, sleepiness, and, finally, unconsciousness.

Should symptoms occur which cannot definitely be identified as either hypoxia or hyperventilation, the following steps should be taken: (1) check your oxygen equipment and put the regulator automix level on 100 percent oxygen (demand or pressure demand system). Continuous flow-check oxygen supply and flow mechanism; (2) after three or four deep breaths of oxygen, the symptoms should improve markedly, if the condition was hypoxia (recovery from hypoxia is rapid); (3) if the symptoms persist, consciously slow your breathing rate until symptoms clear and then resume normal breathing rate. Breathing can be slowed by breathing into a bag or talking loudly.

hypoxia—In simple terms, hypoxia is a lack of sufficient oxygen to keep the brain and other body tissues functioning properly. Wide individual variation occurs with respect to susceptibility to hypoxia. In addition to progressively insufficient oxygen at higher altitudes, anything interfering with the blood's ability to carry oxygen can contribute to hypoxia (anemias, carbon monoxide, and certain drugs). Also, alcohol and various drugs decrease the brain's tolerance to hypoxia.

Your body has no built-in alarm system to let you know when you are not getting enough oxygen. It is impossible to predict when or where hypoxia will occur during a given flight, or how it will manifest itself.

A major early symptom of hypoxia is an increased sense of well-being (referred to as euphoria). This progresses to slow reactions, impaired thinking ability, unusual fatigue, and dull headache feeling.

The symptoms are slow but progressive, insidious in onset, and are most marked at altitudes above 10,000 feet. Night vision, however, can be impaired starting at altitudes as low as 5,000 feet. Heavy smokers may also experience early symptoms of hypoxia at lower altitudes than nonsmokers.

If you observe the general rule of not flying above 10,000 feet without supplemental oxygen, you will not get into trouble.

Hz—Hertz.

I

I in the phonetic alphabet is India (in-dee-ah).

IAS—(*See* AIRSPEED.)

ICAO—International civil aviation organization.

icgic—Icing in clouds.

icgip—Icing in precipitation.

icing, airframe—The effects of ice accretion on aircraft are cumulative—thrust is reduced, drag increases, lift lessens, weight increases. The results are an increase in stall speed and a deterioration of aircraft performance. In extreme cases, 2 to 3 inches of ice can form on the leading edge of the airfoil in less than 5 minutes. It takes but ½ inch of ice to reduce the lifting power of some aircraft by 50 percent and increase the frictional drag by an equal percentage. A pilot can expect icing when flying in visible precipitation such as rain or cloud droplets, and the temperature is 0° centigrade or colder. When icing is detected, a pilot should do one of two things (particularly if the aircraft is not equipped with deicing equipment): he should get out of the area of precipitation, or go to an altitude where the temperature is above freezing. This "warmer" altitude may not always be a lower altitude. Proper pre-flight action includes obtaining information on the freezing level and the above-freezing levels in precipitation areas. Report icing to ATC/FSS.

ident—Identification; also a controller's request for a pilot to activate the aircraft transponder identification feature. This will help the controller to confirm an aircraft identity or to identify an aircraft.

if feasible, reduce speed to (speed)—(*See* SPEED ADJUSTMENT.)

if no transmission received for (time)—Used by ATC in radar approaches to prefix procedures which should be followed by the pilot in event of lost communications.

IFR—Instrument flight rules.

IFR conditions—Are those weather conditions below the minimum for flight under visual flight rules (in a control zone, less than 3 miles visibility and 1,000-foot ceiling).

IFSS—International flight service station.

ILS—Instrument landing system.

imdt—Immediate.

immediately—Used by ATC when such action compliance is required to avoid an imminent situation.

incidence, angle of—Angle between chord of wing and longitudinal axis of airplane.

incr—Increase.

increase speed to (speed)—(*See* SPEED ADJUSTMENT.)

indef—Indefinite.

indicated airspeed—(*See* AIRSPEED.)

indicated altitude—(*See* ALTITUDE.)

in-flight advisory plotting chart—(*See* Figure I-1.)

in-flight failure of radio communication—(*See* EMERGENCY PROCEDURES.)

in-flight weather advisories—(*See* AIRMETS; SIGMETS.)

info—Information.

information request/inreq—A request originated by an FSS for information concerning an overdue VFR aircraft.

initial contact—

1. The term initial contact or initial callup means the first radio call you make to a given facility, or the first call to a different controller or FSS specialist within a facility. Use the following format:

   a. name of facility being called,
   b. your *full* aircraft identification as filed in the flight plan,
   c. type of mesage to follow or your request if it is short, and
   d. the word "Over."

Example: "NEW YORK RADIO, MOONEY THREE ONE ONE ECHO, OVER."

initial contact ***continued***

Example: "COLUMBIA GROUND, CESSNA THREE ONE SIX ZERO FOXTROT, VFR MEMPHIS, OVER."

Example: "MIAMI CENTER BARON FIVE SIX THREE HOTEL, REQUEST VFR TRAFFIC ADVISORIES, OVER."

2. If radio reception is reasonably assured, inclusion of your request, your position or altitude, the phrase "Have numbers" or "Information Charlie received" (for ATIS) in the initial contact helps decrease radio frequency congestion. Use discretion and do not overload the controller with information he does not need. If you do not get a response from the ground station, recheck your radios or use another transmitter but keep the next contact short.

inop—Inoperative.

in-runway lighting—Touchdown zone lighting and runway centerline lighting are installed on some precision approach runways to facilitate landing under adverse visibility conditions. Taxiway turnoff lights may be added to expedite movement of aircraft from the runway.

1. Touchdown Zone Lighting—two rows of transverse light bars disposed symmetrically about the runway centerline in the runway touchdown zone. The system generally extends from 75 to 125 feet of the landing threshold to 3,000 feet down the runway.

2. Runway Centerline Lighting—flush centerline lights spaced at 50-foot intervals beginning 75 feet from the landing threshold and extending to within 75 feet of the opposite end of the runway.

3. Runway Remaining Lighting—is applied to centerline lighting systems in the final 3,000 feet, as viewed from the take-off or approach position. Alternate red and white lights are seen from the 3,000-foot points to the 1,000-foot points, and all red lights are seen for the last 1,000 feet of the runway. From the opposite direction, these lights are seen as white lights.

4. Taxiway turnoff lights—flush lights spaced at 50-foot intervals, defining the curved path of aircraft travel from the runway centerline to a point on the taxiway.

inspection of aircraft—Required every 12 calendar months or, if aircraft is used for hire, every 100 hours.

inst—Instrument.

instrument approach light systems—

1. Instrument approach light systems provide the basic means for transition from instrument flight using electronic approach aids to visual flight and landing. Operational requirements dictate the sophistication and configuration of the approach light system for a particular airport.

2. Condenser-discharge sequenced flashing light systems are installed in conjunction with the instrument approach light system at some airports as a further aid to pilots making instrument approaches. The system consists of a series of brilliant blue-white bursts of light flashing in sequence along the approach lights. It gives the effect of a ball of light traveling toward the runway. An impression of the system as a pilot first observes the flashing lights when making an approach is that of large tracer shells rapidly fired from a point in space toward the runway.

instrument flight experience—Under simulated conditions is part of the training for the private pilot certificate. To obtain an instrument rating certificate, however, the private pilot must meet certain additional standards in ground instruction, flight instruction, and operating skills.

instrument flight experience—(*See* PREREQUISITES FOR PRIVATE PILOT CERTIFICATE.)

Instrument Landing System (ILS)—General:

1. The instrument landing system is designed to provide an approach path for exact alignment and descent of an aircraft on final approach to a runway.

2. The ground equipment consists of two highly directional transmitting systems and, along the approach, three (or fewer) marker beacons. The directional transmitters are known as the localizer and glide slope transmitters.

3. The system may be divided functionally into three parts: guidance information (localizer, glide slope), range information (marker beacon DME), visual information (approach lights, touchdown and centerline lights, runway lights). (*See* Figure I-1.)

instrument rating requirements—To be eligible for an instrument rating, an applicant must

1. Hold a current private pilot certificate with an aircraft rating

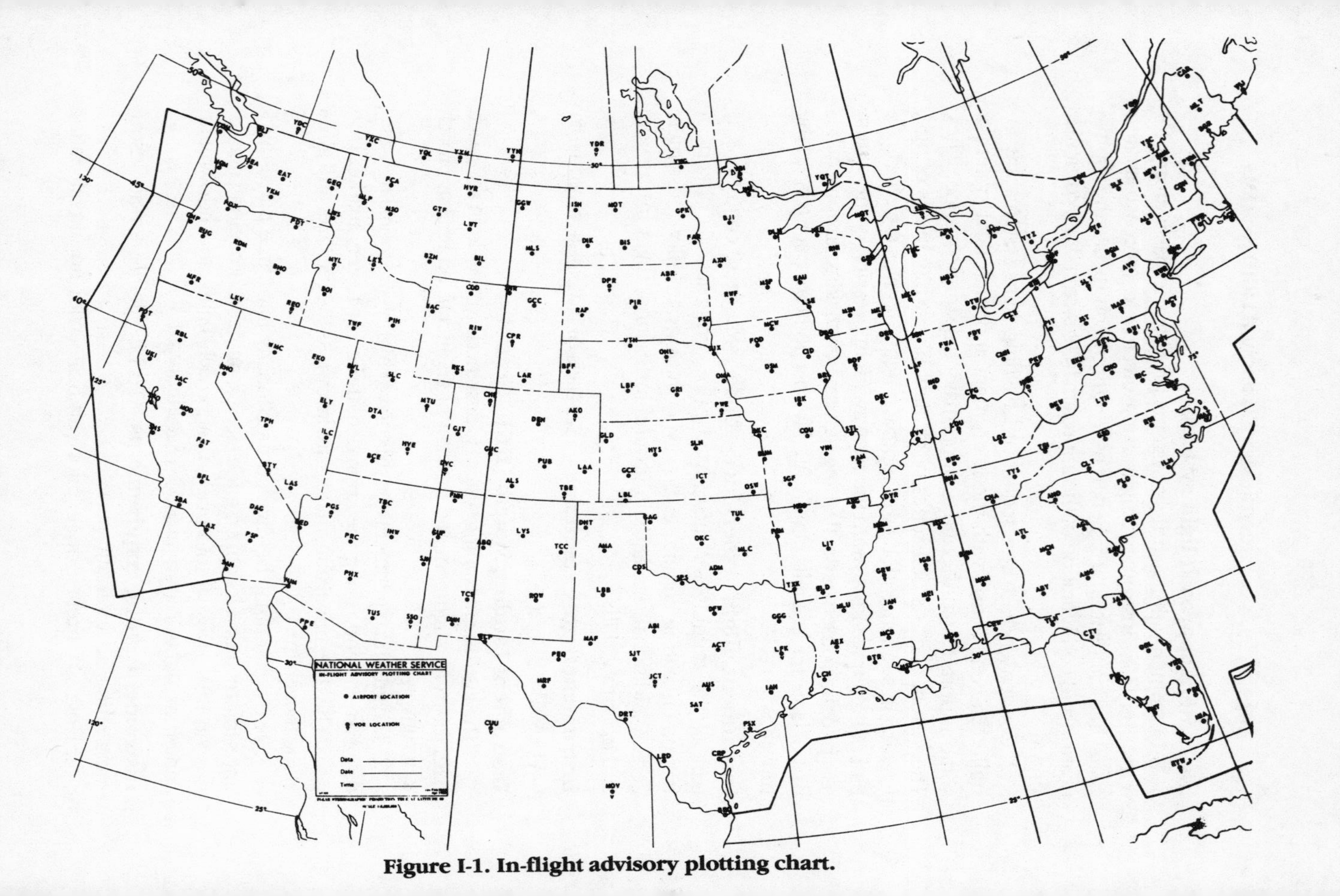

Figure I-1. In-flight advisory plotting chart.

instrument rating requirements *continued*

appropriate to the instrument rating sought.

2. Comply with the requirements for ground, flight, instrument instruction and skill as described in Part 61 of the FAR.

3. Have a total of 125 hours of pilot flight time, of which 50 hours are as pilot in command in cross-country flight in a powered aircraft with other than a student pilot certificate. Each cross-country flight must have a landing at a point more than 50 nautical miles from the original departure point.

4. Pass a written and practical test.

instruments, required—(*See* VISUAL FLIGHT RULES, REQUIRED EQUIPMENT, DAY; VISUAL FLIGHT RULES, REQUIRED EQUIPMENT, NIGHT.)

int—Intersection.

interception procedure—During peacetime, intercepted aircraft will be approached from the stern. Generally two interceptor aircraft will be employed to accomplish the identification. An aircraft which is intercepted by another aircraft shall immediately:

1. Follow the instructions given by the intercepting aircraft, interpreting and responding to the visual signals. (*See* Table I-1.)

2. Notify, if possible, the appropriate air traffic services unit.

3. Attempt to establish radio communication with the intercepting aircraft or with the appropriate intercept control unit, by making a general call on the emergency frequency 243.0 MHz and repeating this call on the emergency frequency 121.5 MHz, if practicable, giving the identity and position of the aircraft and the nature of the flight.

4. If equipped with SSR transponder, select MODE 3/A Code 7700, unless otherwise instructed by the appropriate air traffic services unit. If any instructions received by radio from any sources conflict with those given by the intercepting aircraft by visual or radio signals, the intercepted aircraft shall request immediate clarification while continuing to comply with the instructions given by the intercepting aircraft.

International Flight Information Manual (IFIM)—A publication designed primarily as a pilot's preflight planning guide for flights into foreign airspace and for flights returning to the U.S. from foreign locations.

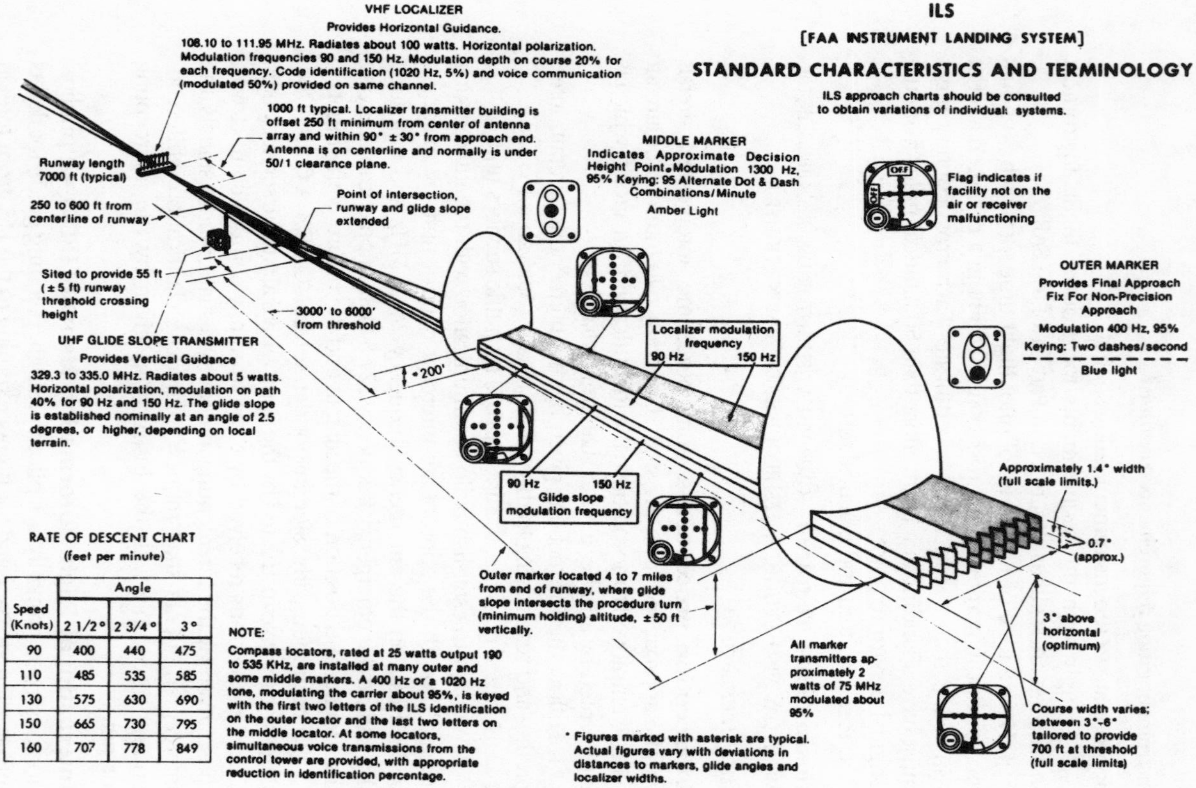

ILS

[FAA INSTRUMENT LANDING SYSTEM]

STANDARD CHARACTERISTICS AND TERMINOLOGY

ILS approach charts should be consulted to obtain variations of individual systems.

VHF LOCALIZER

Provides Horizontal Guidance.

108.10 to 111.95 MHz. Radiates about 100 watts. Horizontal polarization. Modulation frequencies 90 and 150 Hz. Modulation depth on course 20% for each frequency. Code identification (1020 Hz, 5%) and voice communication (modulated 50%) provided on same channel.

1000 ft typical. Localizer transmitter building is offset 250 ft minimum from center of antenna array and within 90° ± 30° from approach end. Antenna is on centerline and normally is under 50/1 clearance plane.

Runway length 7000 ft (typical)

250 to 600 ft from centerline of runway

Point of intersection, runway and glide slope extended

Sited to provide 55 ft (± 5 ft) runway threshold crossing height

3000' to 6000' from threshold

*200'

UHF GLIDE SLOPE TRANSMITTER

Provides Vertical Guidance

329.3 to 335.0 MHz. Radiates about 5 watts. Horizontal polarization, modulation on path 40% for 90 Hz and 150 Hz. The glide slope is established nominally at an angle of 2.5 degrees, or higher, depending on local terrain.

MIDDLE MARKER

Indicates Approximate Decision Height Point. Modulation 1300 Hz, 95% Keying: 95 Alternate Dot & Dash Combinations/Minute

Amber Light

Flag indicates if facility not on the air or receiver malfunctioning

OFF

OUTER MARKER

Provides Final Approach Fix For Non-Precision Approach

Modulation 400 Hz, 95%

Keying: Two dashes/second

Blue light

Localizer modulation frequency
90 Hz 150 Hz

90 Hz 150 Hz
Glide slope modulation frequency

Approximately 1.4° width (full scale limits.)

0.7° (approx.)

3° above horizontal (optimum)

Course width varies; between 3°-6° tailored to provide 700 ft at threshold (full scale limits)

Outer marker located 4 to 7 miles from end of runway, where glide slope intersects the procedure turn (minimum holding) altitude, ± 50 ft vertically.

All marker transmitters approximately 2 watts of 75 MHz modulated about 95%

RATE OF DESCENT CHART
(feet per minute)

Speed (Knots)	Angle 2 1/2°	2 3/4°	3°
90	400	440	475
110	485	535	585
130	575	630	690
150	665	730	795
160	707	778	849

NOTE:

Compass locators, rated at 25 watts output 190 to 535 KHz, are installed at many outer and some middle markers. A 400 Hz or a 1020 Hz tone, modulating the carrier about 95%, is keyed with the first two letters of the ILS identification on the outer locator and the last two letters on the middle locator. At some locators, simultaneous voice transmissions from the control tower are provided, with appropriate reduction in identification percentage.

* Figures marked with asterisk are typical. Actual figures vary with deviations in distances to markers, glide angles and localizer widths.

Figure I-2. Instrument landing system.

interrogator—The ground-based surveillance radar beacon transmitter-receiver which normally scans in synchronism with a primary radar transmitting discrete radio signals which repetitiously requests all transponders, on the mode being used, to reply. The replies received are mixed with the primary radar returns and displayed on the same plan position indicator (radar scope). Also applied to the airborne element of the TACAN/DME system. (*See* TRANSPONDER.)

intersection—

1. A point defined by any combination of courses, radials, or bearings of two or more navigational aids.

2. Used to describe the point where two runways cross, a taxiway and a runway cross, or two taxiways cross.

intersection takeoffs—

1. In order to enhance airport capacities, reduce taxiing distances, minimize departure delays, and provide for more efficient movement of air traffic, controllers may initiate intersection takeoffs as well as approve them when the pilot requests. If for ANY reason a pilot prefers to use a different intersection or the full length of the runway, or desires to obtain the distance between the intersection and the runway end, HE IS EXPECTED TO INFORM ATC ACCORDINGLY.

2. Controllers are required to separate small propeller driven aircraft (less than 12,500 pounds) taking off from an intersection on the same runway following a large aircraft by ensuring that at least a 3-minute interval exists between the time that the preceding turbojet has taken off and the succeeding aircraft begins take-off roll. To inform the pilot of the required 3-minute hold, the controller will state, "Hold for wake turbulence." If after considering wake turbulence hazards, the pilot feels that a lesser time interval is appropriate, he may request a waiver to the 3-minute interval. Pilots must initiate such a request by stating, "Request waiver to 3-minute interval," or by making a similar statement. Controllers may then issue a takeoff clearance if other traffic permits, since the pilot has accepted responsibility for this own wake turbulence separation.

3. The 3-minute interval will not be applied when the intersection is 500 feet or less from the threshold and both aircraft are taking off in the same direction. When a small aircraft is departing from an intersection 500 feet or less from the runway threshold behind a large nonheavy aircraft and both aircraft are taking off in the same direction,

Table I-1.

INTERCEPTION SIGNALS

SIGNALS INITIATED BY INTERCEPTED AIRCRAFT AND RESPONSES BY INTERCEPTING AIRCRAFT

Series	INTERCEPTED Aircraft Signals	Meaning	INTERCEPTING Aircraft Responds	Meaning
1	DAY—Rocking wings from a position in front and, normally, to the left of intercepted aircraft and, after acknowledgement, a slow level turn, normally to the left, on to the desired course.	You have been intercepted. Follow me.	DAY—Rocking wings and following.	Understood, will comply.
	NIGHT—Same and, in addition, flashing navigational and, if available, landing lights at irregular intervals.		NIGHT—Same and, in addition, flashing navigational and, if available, landing lights at irregular intervals.	
	Note: Meteorological conditions or terrain may require the intercepting aircraft to take up a position in front and to the right of the intercepted aircraft and to make the subsequent turn to the right.			
2	DAY OR NIGHT—An abrupt break-away maneuver from the intercepted aircraft consisting of a climbing turn of 90 degrees or more without crossing the line of flight of the intercepted aircraft.	You may proceed.	DAY OR NIGHT—Rocking wings.	Understood, will comply.

SIGNALS INITIATED BY INTERCEPTED AIRCRAFT AND RESPONSES BY INTERCEPTING AIRCRAFT

Series	INTERCEPTED Aircraft Signals	Meaning	INTERCEPTING Aircraft Responds	Meaning
3	DAY—Circling aerodrome, lowering landing gear and overflying runway in direction of landing.	Land at this aerodrome.	DAY—Lowering landing gear, following the intercepting aircraft and, if after overflying the runway landing is considered safe, proceeding to land.	Understood, will comply.
	NIGHT—Same and, in addition, showing steady landing lights.		NIGHT—Same and, in addition, showing steady landing lights (if carried).	
4	DAY—Raising landing gear while passing over landing runway at a height exceeding 300m (1,000 ft) but not exceeding 600m (2,000 ft) above the aerodrome.	Aerodrome you have designated is inadequate.	DAY OR NIGHT—If it is desired that the intercepted aircraft follow the intercepting aircraft to an alternate aerodrome, the intercepting aircraft raises its landing gear and uses the Series 1 signals prescribed for intercepting aircraft.	Understood, follow me.
	NIGHT—Flashing landing lights while passing over landing runway at a height exceeding 300m (1,000 ft) but not exceeding 600m (2,000 ft) above the aerodrome level, and continuing to circle the aerodrome. If unable to flash landing lights, flash any other lights available.		If it is decided to release the intercepted aircraft, the intercepting aircraft uses the Series 2 signals prescribed for intercepting aircraft.	Understood, you may proceed.

intersection takeoffs *continued*

the 3-minute interval will not be applied. Controllers will issue a clearance to permit the small aircraft to alter course to avoid the flight path of the preceding departure.

4. The 3-minute interval is always mandatory behind a heavy aircraft.

interstate air commerce—Carriage by aircraft of persons or property for compensation or hire; operation of aircraft for business or vocation between states.

ints—Intensity.

intsfy—Intensify.

intl—International.

I say again—Pilot/controller phraseology for the message will be repeated.

ISJTA—Intensive student jet training area.

island reporting service—(*See* LAKE/ISLAND, MOUNTAIN, AND SWAMP REPORTING SERVICE.)

isold—Isolated.

J

J in the phonetic alphabet is Juliett (jew-lee-ett).

jamming—Electronic or mechanical interference which may disrupt the display of aircraft on radar or the transmission/reception of radio communications/navigation.

jet advisory areas—VFR Operation, jet advisory areas have not been designated positive control airspace, and VFR flight is permitted within these areas under the following conditions.

1. Radar jet advisory area—Prior approval from ATC is required unless the aircraft has a functioning transponder and is operated on the appropriate code.

2. Nonradar jet advisory areas—Prior approval from ATC is required in all cases. NOTE: Jet advisory areas and the flight levels comprising

these areas are depicted on en route-high altitude charts.

jet blast—Jet engine exhaust (thrust stream turbulence). (*See* WAKE TURBULENCE.)

jet route—A route designed to serve aircraft operations from 18,000 feet MSL up to and including flight level 450. The routes, referred to as "J" routes, are numbered to identify the designated route, e.g. J 105.

jtstr—Jet stream.

K

K in the phonetic alphabet is Kilo (kee-lo).

K—A symbol for an air mass colder than the surface over which it is running.

"Keep-'Em-High"—A Federal Aviation Administration program in which arriving high performance (turbojet and turboprop) aircraft will be kept at the highest possible altitude for as long as possible, while departing high performance aircraft will be climbed to the highest possible altitude filed by the pilot as soon as possible.

kHz—Kilohertz (1,000 hertz).

kilometer—1,000 meters (3,280.8 feet or about ⅝ of a mile).

kinesthesia—The sense which detects and estimates motion without reference to vision or hearing.

knot—Unit of speed equaling one nautical mile per hour.

known traffic—With respect to ATC clearances, means aircraft whose altitude, position, and intentions are known to ATC.

Koch chart—An aid to the pilot in figuring the effect of airport temperature and altitude on his take-off distance and rate of climb.

KT—Knot(s).

L

L in the phonetic alphabet is Lima (lee-mah).

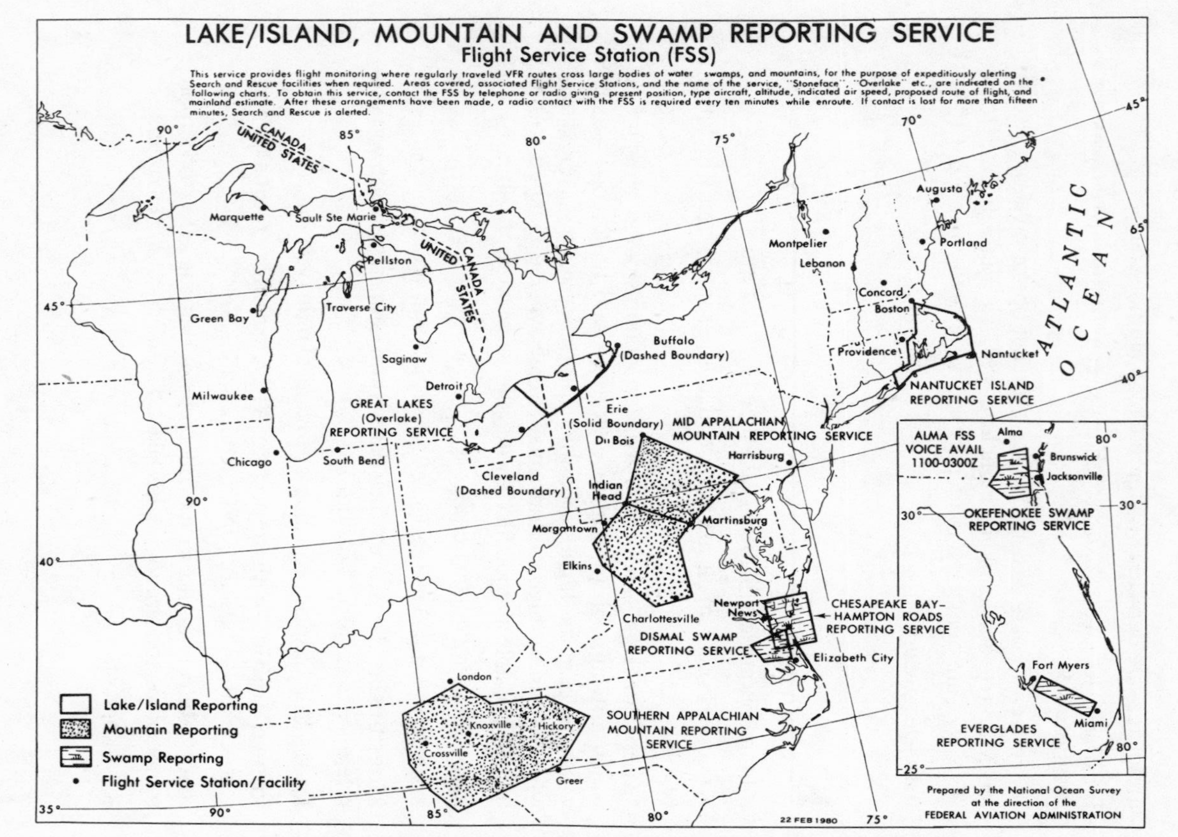

Figure L-1. Lake/island, mountain, and swamp reporting service.

Lake/Island, Mountain and Swamp Reporting Service—(*See* Figure L-1.)

land plane—An airplane designed to rise from and alight on the ground.

landing—The act of terminating flight and bringing the airplane to rest; used both for land and seaplanes.

landing area—Any area suitable for the landing of an airplane.

landing clearance—Is required from air traffic control before landing at any airport with a tower.

landing direction indicator—A tetrahedron or a tee installed when conditions at the airport warrant its use, located at the center of the segmented circle, and used to indicate the direction in which landings and take-offs should be made. The large end (crossbar) of a tee is in the direction of landing. The small end of a tetrahedron points in the direction of landing. (*See* Figures L-2 and L-3.) Pilots are cautioned against using a tetrahedron for any purpose other than as an indicator of landing direction, and to disregard the tetrahedron at an airport with an operating tower. Tower instructions supersede tetrahedron indications.

landing gear—The understructure which supports the weight of the airplane while at rest.

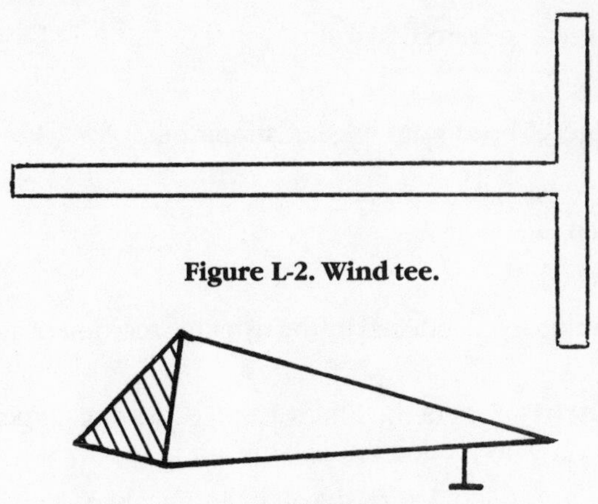

Figure L-2. Wind tee.

Figure L-3. Tetrahedron

landing light—Is required on any aircraft, operated for hire, which is landing at night.

landing roll—The distance from the point of touchdown to the point where the aircraft can be brought to a stop or exit the runway.

landing sequence—The order in which aircraft are positioned for landing. (*See* APPROACH SEQUENCE.)

landing strip indicators—Installed in pairs (as shown in Figure 5-1., Segmented Circle, on page 168) and used to show the alignment of landing strips.

landings, simultaneous—(*See* SIMULTANEOUS LANDINGS ON INTERSECTING RUNWAYS.)

last assigned altitude—The last altitude/flight level assigned by ATC and acknowledged by the pilot. (*See* MAINTAIN.)

lat—Latitude.

lateral separation—The lateral spacing of aircraft at the same altitude by requiring operation on different routes or in different geographical locations. (*See* SEPARATION.)

LC—Local telephone number (local call).

lctd—Located.

lctn—Location.

LDA—Localizer-type directional aid.

ldg—Landing.

leading edge—The forward edge of any airfoil.

lgt—Light.

lgtd—Lighted.

lgts—Lights.

lift—Supporting force induced by the dynamic reaction of air against wing.

lift component—Sum of the forces acting on a wing perpendicular to the direction of its motion through the air.

light gun—A handheld directional light signaling device which emits a brilliant narrow beam of white, green, or red light as selected by the tower controller. The color and type of light transmitted can be used to approve or disapprove anticipated pilot actions where radio communications is not available. The light gun is used for controlling traffic operating in the vicinity of the airport and on the airport movement area.

light signals—

1. The following procedures are used by airport traffic control towers to control aircraft not equipped with radio. (These same procedures will be used to control radio-equipped aircraft if radio contact cannot be established.) A directive light signal which emits an intense, narrow beam of a selected color (red, white or green) offers the advantage of some control, but pilots should be cognizant of the disadvantages inherent in this method:

a. The pilot may not be looking at the control tower at the time a signal is directed toward him.

b. The directions transmitted by a light signal are very limited, since only approval or disapproval of a pilot's anticipated actions may be transmitted. No supplemental or explanatory information may be transmitted except by the use of the "general warning signal" which advises the pilot to be on the alert.

2. Portable traffic control light signals:

Color and Type	On the Ground	In Flight
STEADY GREEN	cleared for take-off	cleared to land
FLASHING GREEN	cleared to taxi	return for landing (to be followed by steady green at proper time)
STEADY RED	stop	give way to other aircraft and continue circling
FLASHING RED	taxi clear of landing area (runway) in use	airport unsafe, do not land
FLASHING WHITE	return to starting point on airport	
ALTERNATING RED AND GREEN	General Warning Signal—Exercise Extreme Caution	

3. Between sunset and sunrise, a pilot wishing to attract the atten-

light signals *continued*

tion of the control tower should turn on a landing light and taxi the aircraft in position so that light is visible to the tower. The landing light should remain on until appropriate signals are received from the tower.

4. Pilots should acknowledge light signals by moving the ailerons or rudder during the hours of daylight, or by blinking the landing or navigation lights during the hours of darkness.

5. In control zones, operation of the airport beacon during the hours of daylight often indicates that the ground visibility is less than 3 miles and/or the ceiling is less than 1,000 feet. ATC clearance in accordance with FAR Part 91 would be required for landing, takeoff, and flight in the traffic pattern. Pilots should not rely solely on the operation of the airport beacon to indicate weather conditions, IFR versus VFR. At locations with control towers and if controls are provided, ATC personnel turn the beacon on. However, at many airports throughout the country, the airport beacon is turned on by a photoelectric cell or time clocks and ATC personnel have no control as to when it shall be turned on. Also, there is no regulatory requirement for daylight operation and pilots are reminded that it remains their responsibility for complying with proper preflight planning in accordance with FAR Part 91.

lights, aircraft—From sunset to sunrise, aircraft position lights must be visible during operation. As the pilot is seated in the aircraft, his right wing carries a green light, his left a red light, the tail a white light.

lights, auxiliary—

1. The auxiliary lights are of two general kinds: code beacons and course lights. The code beacon, which can be seen from all directions, is used to identify airports and landmarks and to mark hazards. The number of code beacon flashes are:
   a. Green coded flashes not exceeding 40 flashes or character elements per minute, or constant flashes 12 to 15 per minute, for identifying land airports;
   b. Yellow coded flashes not exceeding 40 flashes or character elements per minute, or constant flashes 12 to 15 per minute, for identifying water airports;

c. Red flashes, a constant rate, 12 to 40 flashes per minute, for marking hazards.

2. The course light, which can be seen clearly from only one direction, is used only with rotating beacons of the federal airway system; two course lights, back to back, direct coded flashing beams of light in either direction along the course of airway.

lights, military airports—Military airport beacons flash alternately white and green, but are differentiated from civil beacons by dual-peaked (two quick) white flashes between the green flashes.

light systems, control of—

1. Operation of approach light systems and runway lighting is controlled by the control tower. At some locations the FSS may control the lights where there is no control tower in operation.
2. Pilots may request that lights be turned on, off, or that the intensity be increased or decreased. The HIRL, MIRL, and the instrument approach lights have intensity controls which can be adjusted to meet pilot's requests. Some sequence flashing lights also have intensity controls, but can be turned on or off at pilot's request. (*See* APPROACH LIGHT SYSTEM, RUNWAY EDGE LIGHT SYSTEMS, RUNWAY END IDENTIFIER LIGHTS, VISUAL APPROACH SLOPE INDICATOR [VASI].)

light systems, runway edge—Runway edge lights are used to outline the edges of runways during periods of darkness and restricted visibility conditions. These light systems are classified according to the intensity or brightness they are capable of producing: High Intensity Runway Lights (HIRL), Medium Intensity Runway Lights (MIRL), and Low Intensity Runway Lights (LIRL). The HIRL and MIRL systems have variable intensity controls, whereas the LIRLs normally have one intensity setting.

The runway edge lights are white except that on instrument runways aviation yellow replaces white on the last 2,000 feet or half the runway length, whichever is less, to form a caution zone for landings. The lights marking the longitudinal limits of the runway emit red light toward the runway to indicate the end of runway to a departing aircraft and emit green outward from the runway end to indicate the threshold to landing aircraft.

line of sight—Possible at average altitudes: 1,000 feet—45 miles; 5,000 feet—100 miles; 10,000 feet—140 miles.

liquor—(*See* ALCOHOL.)

LIRL—Low intensity runway lights.

LMM—Compass locator at middle marker ILS.

lmt—Limit.

load—The forces acting on a structure. These may be static (as with gravity) or dynamic (as with centrifugal force), or a combination of static and dynamic.

load factor—The sum of the loads on a structure, including the static and dynamic loads, expressed in "G" units.

loc—Localizer.

local traffic—Aircraft operating in the traffic pattern or within sight of the tower, or aircraft known to be departing or arriving from flight in local practice areas, or aircraft executing practice instrument approaches at the airport. (*See* TRAFFIC PATTERN.)

locator beacons—

1. Locator beacons of various types, which are independently powered, reliable, and of incalculable value in an emergency, have been developed and are gaining wide acceptance as a means of locating downed aircraft and their occupants. These electronic, battery operated beacons are not a fire hazard. They are designed to emit a distinctive downward-swept audio tone for homing purposes on 121.5 MHz and/or 243 MHz, preferably on both emergency frequencies. The power source should be capable of providing power for continuous operation from 24 to 48 hours or more at a very wide range of ambient temperatures and can expedite search and rescue operations as well as facilitate accident investigation and analysis.

2. This equipment is ideally suited for general aviation and small private aircraft, and these beacons are made by several electronic manufacturers for civil aviation use. For purposes of discussion, locator beacons can be broadly classified into three basic types: *Crash Locator Beacon* applies to the device which is installed in or onto the structure of the aircraft and which may be ejectable; *Personnel Locator Beacon* describes the relatively small, portable device which can be

carried as a part of the personal equipment of the flight crew or occupants; *Survival Radio Equipment* refers to the device which is usually carried as a part of the internationally agreed upon emergency equipment for extended overwater operations and is required for certain air carrier operations. (*Also see* individual entries for these locator devices.)

3. Once the beacon has been activated and the signal detected, it will be a simple matter for the search aircraft with homing equipment to locate the scene. Search patterns have been developed that enable search aircraft equipped with but one receiver to locate the locator beacon site.

4. Some models of locator beacons may be of value for transmitting a signal for inflight emergencies. If an emergency situation occurs in conjunction with a radio failure, the pilot could actuate his beacon signal to supplement other actions taken to declare the emergency. DF stations hearing the signal would take bearings on the aircraft in distress and notify SAR.

5. Pilots of aircraft equipped with locator beacons are encouraged to include this information in the REMARKS portion of their flight plans.

6. In addition to depleting the batteries, accidental triggering of the beacon or improper test procedures could cause an unnecessary search. Check the off/on switch prior to and after each flight, and store the beacon in a secure place until needed.

log—Flight-by-flight record of all operations of an airplane, engine, or pilot, listing flight time, area of operation, and other pertinent information.

logbooks—Flight time used to meet experience requirements for pilot ratings must be shown by a reliable record.

logs, engine—Must be maintained by owners of aircraft.

LOM—Compass locator at outer marker ILS.

long—Longitude.

longeron—The principal longitudinal structural member in a fuselage.

Long Island Sound Reporting Service (LIRS)—The New York and Windsor Locks FSSs provide Long Island Sound reporting service

Long Island Sound Reporting Service (LISRS) *continued*

on request for aircraft traversing Long Island Sound. When requesting the service pilots should ask for SOUND REPORTING SERVICE and should be prepared to provide the following appropriate information:

1. Type and color of aircraft.
2. The specific route and altitude across the sound, including the shore crossing point.
3. The overwater crossing time.
4. Number of persons on board.
5. True air speed.

Radio contacts are desired at least every 10 minutes; however, for flights of shorter duration a mid-sound report is requested. If contact is lost for more than 15 minutes, search and rescue will be alerted. Pilots are responsible for canceling their request for the Long Island Sound reporting service when outside the service area boundary. Aircraft experiencing radio failure will be expected to land as soon as practicable and to cancel their request for the service.

6. COMMUNICATIONS: Primary communications—pilots are to transmit on 122.1 MHz and listen on one of the following VOR frequencies:
   a. NEW YORK FSS CONTROLS:
      i. Hampton VORTAC (FSS transmits on 113.6 and receives on 122.1 MHz).
      ii. Calverton VORTAC (FSS transmits on 117.2 and receives on standard FSS frequencies).
      iii. Kennedy VORTAC (FSS transmits on 115.9 and receives on 122.1 MHz).
   b. WINDSOR LOCKS FSS CONTROLS:
      i. Madison VORTAC (FSS transmits on 110.4 and receives on 122.1 MHz).
      ii. Trumbull VOR (FSS transmits on 111.8 and receives on 122.1 MHz).
      iii. Bridgeport VOR (FSS transmits on 108.8 and receives on 122.1 MHz).

longitudinal separation—The longitudinal spacing of aircraft at the same altitude by a minimum distance expressed in units of time or miles. (*See* SEPARATION.)

lost—If you are not sure of your position, call any nearby Flight Service Station, at the radio frequency shown on your sectional chart. If unsuccessful there, transmit on 121.5 MHz the emergency frequency, your type of aircraft, call number, altitude, and heading and ask for direction assistance. If both your radio transmitter and receiver are inoperative, or only your transmitter is functioning, squawk Code 7700 if transponder equipped.

Radar facilities are equipped so that Code 7700 normally triggers an alarm or special indicator at all control positions. Pilots should understand that they might not be within a radar coverage area. Therefore, they should continue squawking Code 7700 and establish radio communications as soon as possible.

lost communications/two-way radio communications failure—Loss of the ability to communicate by radio. Aircraft are sometimes referred to as NORDO (No Radio). Standard pilot procedures are specified in FAR Part 91. Radar controllers issue procedures for pilots to follow in the event of lost communications during a radar approach when weather reports indicate that an aircraft will likely encounter IFR weather conditions during the approach.

low altitude airway structure/federal airways—The network of airways serving aircraft operations up to but not including 18,000 feet MSL. (*See* AIRWAY.)

low altitude alert, check your altitude immediately—(*See* SAFETY ADVISORY.)

low altitude military training routes—(*See* MILITARY TRAINING ROUTES.)

low approach—

1. A low approach (sometimes referred to as a low pass) is the go-around maneuver following an approach. Instead of landing or making a touch-and-go, a pilot may wish to go around (low approach) in order to expedite a particualr operation; a series of practice instrument approaches is an example of such an operation. Unless otherwise authorized by ATC, the low approach should be made straight ahead, with no turns, or climb made until the pilot has made a thorough visual check for other aircraft in the area.

2. When operating within an airport traffic area, a pilot intending to

low approach *continued*

make a low approach should contact the tower for approval. This request should be made prior to starting the final approach.

3. When operating to an airport not within an airport traffic area; a pilot intending to make a low approach should, prior to leaving the final approach fix inbound, so advise the FSS, UNICOM, or make a broadcast, as appropriate.

low frequency (LF)—The frequency band between 30 and 300 kHz. (Refer to AIM.)

low-level wind shear alert system (LLWSAS)—This computerized system detects the presence of a possible hazardous low-level wind shear by continuously comparing the winds measured by sensors installed around the periphery of an airport with the wind measured at the centerfield location. If the difference between the centerfield wind sensor and a peripheral wind sensor becomes excessive, a thunderstorm or thunderstorm gust front wind shear is probable. When this condition exists, the tower controller will provide arrival and departure aircraft with an advisory of the situation which includes the centerfield wind plus the remote site location and wind.

low/medium frequency (L/MF) radio range—These ranges are classified by their type of antennae. Two types of low-frequency ranges are in use: loop range (L) and adcock range (A). Low-frequency radio range courses are subject to disturbances resulting in multiple courses, signal fades, and surges over rough country. Pilots flying over unfamiliar routes are cautioned to be on the alert to detect these vagaries, particularly over mountainous terrain.

LRCO—Limited remote communications outlet.

lubber line—The small reference line used in reading the figures from the card of an aeronautical compass.

lvl—Level.

lwr—Lower.

lyr—Layer.

M

M in the phonetic alphabet is Mike (mike).

m—Symbol for maritime air mass.

MAA—Maximum authorized altitude.

mag—Magnetic.

magenta-tinted bands on aeronautical charts—(*See* Figure A-8.)

magnetic compass—(*See* COMPASS, MAGNETIC.)

magnetic dip—The tendency of the magnetic compass to point down as well as north in certain latitudes, causing a northerly turn error and an acceleration error.

maint—Maintain, maintenance.

maintain—

1. Concerning altitude/flight level, the controller's term means to remain at the altitude/flight level specified. The phrase "climb and" or "descend and" normally precedes "maintain" and the altitude assignment, e.g., "descend and maintain 5,000."

2. Concerning other ATC instructions, the term is used in its literal sense, e.g., maintain VFR.

maintenance records—Of prescribed data, are required for each aircraft and each engine used in civil aviation.

make short approach—Used by Air Traffic Control to inform a pilot to alter his traffic pattern so as to make a short final approach. (*See* TRAFFIC PATTERN.)

maneuver—Any planned motion of an airplane, whether in the air or on the ground.

maneuvering speed—The maximum speed at which the flight controls can be fully deflected without damage to the aircraft structure. It may be found in the airplane's flight manual, and is useful for guidance in performing flight maneuvers, or normal operations in severe turbulence.

manifold pressure—Means absolute pressure as measured at the appropriate point in the induction system; usually expressed in inches of mercury.

marker beacons—Serve to identify a particular location in space along an airway or on the approach to an instrument runway. This is done by means of a 75-MHz transmitter which transmits a directional signal to be received by aircraft flying overhead. These markers are generally used in conjunction with low frequency radio ranges and the instrument landing system as point designators. Three classes of en route markers are now in general use: Fan Marker (FM); Low Powered Fan Marker (LFM); and Z Marker. They transmit the letter "R" (dot dash dot) identification, or (if additional markers are in the same area) the letter "K," "P," "X," or "Z," respectively. (*Also see* INSTRUMENT LANDING SYSTEM.)

max—Maximum.

maximum authorized altitude (MAA)—The highest altitude on a federal airway, jet route, or other direct route for which an MEA is designated in Federal Aviation Regulations, Part 95, at which adequate reception of navigation aid signals is assured.

mayday—Taken from the French *M'aidez* meaning "help me," this is the international radiotelephone distress signal. When repeated three times, it indicates imminent and grave danger and that immediate assistance is requested.

MCA—Minimum crossing altitudes.

mdt—Moderate.

MEA—Minimum en route IFR altitude.

med—Medium.

medical certificates—Acceptable evidence of physical fitness on a form prescribed by the FAA; required for all pilots of civil aircraft;
first class—Required for airline transport pilot rating. Good for 6 months if scheduled for air carrier or commuter air taxi; otherwise 1 year;
second class—For private and student pilots; expires the 24th month after the month in which issued. For commercial pilots, expires the 6th month after the month in which issued;
third class—Required for student and private pilots; expires the 24th month after the month in which issued.

medical examiners, approved by Federal Aviation Administration—Names and addresses in each local area available from nearest Federal Aviation Administration office. (See telephone book under United States Government, Federal Aviation Administration.)

meml—Memorial.

meteorology—The study of weather; recognizing dangerous weather conditions and evaluating weather reports are part of the questions on the examination for private pilot's license.

metering—A method of time regulating arrival traffic flow into a terminal area so as not to exceed a predetermined terminal acceptance rate.

metering fix—A fix along an established route from over which aircraft will be metered prior to entering terminal airspace. Normally, this fix should be established at a distance from the airport which will facilitate a profile descent 10,000 feet above airport elevation (AAE) or above.

Mexico—(*See* FLIGHT OUTSIDE THE UNITED STATES.)

MHz—Megahertz (1 million cycles per second).

mi—Mile.

microphone technique—
1. Proper microphone technique is important in radiotelephone communications. Transmissions should be concise and in normal conversational tone. NOTE: Identification of Aircraft. Pilots are requested to exercise care that the identification of their aircraft is

microphone technique *continued*

clearly transmitted in each contact with an ATC facility. Also, pilots should be certain that their aircraft are clearly identified in ATC transmissions before taking action on an ATC clearance.

2. If you are attempting to establish contact with a ground station and you are receiving on a different frequency than that transmitted, indicate the VOR name or the frequency on which you expect a reply. Most FSSs and control facilities can transmit on several VOR stations in the area. Use the appropriate FSS call sign as indicated on charts.

Example: New York FSS transmits on the Kennedy, Hampton and Calverton VORTACs. If you are in the Calverton area, your callup should be "NEW YORK RADIO, CESSNA THREE ONE SIX ZERO FOXTROT, RECEIVING CALVERTON VOR, OVER."

If the chart indicates FSS frequencies above the VORTAC or in FSS communications boxes, transmit or receive on those frequencies nearest your location.

When unable to establish contact and you wish to call *any* ground station, use the phrase "ANY RADIO (tower) (station), GIVE CESSNA THREE ONE SIX ZERO FOXTROT A CALL ON (frequency) OR (VOR)." If an emergency exists or you need assistance, so state.

3. Keep your contacts as brief as possible. Pilots should not read back altimeter setting, taxi instructions, or wind and runway information to towers except for verification or clarification of instructions. Other pilots are waiting to use the channel.

4. Contact the nearest Flight Service Station. Don't continually attempt to see how far your transmitter will reach. If in doubt about the frequency for contacting an FSS, transmit on 122.1 MHz and advise them of the frequency you are listening on.

5. Avoid calling stations at 15 minutes past the hour, because of interference with scheduled weather broadcasts.

6. When making a position report, pilots should in all cases state the name of the reporting point over which, or in relation to which, they are reporting. The phrase "OVER YOUR STATION" should not be used.

middle ear discomfort—(*See* EAR, MIDDLE, DISCOMFORT OR PAIN.)

military airports—(*See* LIGHTS, MILITARY AIRPORTS.)

military climb corridors—Are restricted areas shown on aeronautical charts. Civil aircraft may not fly through these areas without prior approval from the controlling agency.

military fields, civil use of—United States Army, Air Force, Navy, and Coast Guard fields are open to civil fliers only in emergency unless prior permission is granted.

military fields, heavy traffic around—Pilots are advised to exercise vigilance when in close proximity to most military airports, as these may have jet aircraft traffic patterns extending up to 2,500 feet above the surface. In addition, they may have an unusually heavy concentration of jet aircraft operating within a 25 nautical mile radius and from the surface to all altitudes. This precautionary note also applies to the larger civil airports.

military training routes (MTR)—The MTRs program is a joint venture by the FAA and the Department of Defense (DOD). MTR routes are mutually developed for use by the military for the purpose of conducting low-altitude, high-speed training. The routes above 1,500 feet above ground level (AGL) are developed to be flown, to the maximum extent possible, under IFR. The routes at 1,500 feet AGL and below are generally developed to be flown under Visual Flight Rules (VFR).

Generally, MTRs are established below 10,000 feet MSL for operations at speeds in excess of 250 knots. However, route segments may be defined at higher altitudes for purposes of route continuity. For example, route segments may be defined for descent, climbout, and mountainous terrain. There are IFR and VFR routes as follows:

1. IFR Military Training Routes—IR: Operations on these routes are conducted in accordance with IFRs regardless of weather conditions.
2. VFR Military Training Routes—VR: Operations on these routes are conducted in accordance with VFRs.

Routes are identified by gray lines on sectional charts as follows:

1. *IR and VR at or below 1,500 feet AGL* (with no segment above 1,500) will be identified by four digit numbers; e.g., IR 1006, VR 1007, etc.
2. *IR and VR above 1,500 feet AGL* (segments of these routes may be below 1,500) will be identified by three digit numbers; e.g., IR 008, VR 009, etc.

military training routes (MTR) *continued*

Nonparticipating aircraft are not prohibited from flying within an MTR; however, extreme vigilance should be exercised when conducting flight through or near these routes. Pilots should contact FSSs within 100 NM of a particular MTR to obtain current information or route usage in their vicinity. Information available includes times of scheduled activity, altitudes in use on each route segment, and actual route width. Route width varies for each MTR and can extend several miles on either side of the charted MTR centerline.

min—Minimum or minute.

minimum age, student pilot—16 years of age. (*See* PRIVATE PILOT, PREREQUISITES FOR CERTIFICATE.)

minimum age, private pilot—17 years of age. (*See* PRIVATE PILOT, PREREQUISITES FOR CERTIFICATE.)

minimum crossing altitudes (MCA)—The lowest altitudes at certain radio fixes at which an aircraft must cross when proceeding in the direction of a higher minimum en route IFR altitude.

minimum en route IFR altitude (MEA)—The altitude in effect between radio fixes which assures acceptable navigational signal coverage and meets obstruction clearance requirements between those fixes.

minimum fuel—Indicates that an aircraft's fuel supply has reached a state where, upon reaching the destination, it can accept little or no delay. This is not an emergency situation but merely indicates an emergency situation is possible should any undue delay occur.

minimum holding altitude (MHA)—The lowest altitude prescribed for a holding pattern which assures navigational signal coverage, communications, and meets obstacle clearance requirements.

minimum obstruction clearance altitude (MOCA)—The specified altitude in effect between radio fixes on VOR airways, off-airway routes or route segments, which meets obstruction clearance requirements for the entire route segment and which assures acceptable navigational signal coverage only within 25 statute miles of a VOR.

minimum reception altitude (MRA)—The lowest altitude required to receive adequate signals to determine specific VOR/VORTAC/TACAN fixes.

minimum safe altitude—Over congested areas, 1,000 feet above highest obstacle within a 2,000 foot horizontal radius of the aircraft (or high enough to glide clear in case of power failure); not over congested areas, 500 feet above surface or structure.

minimum safe altitude warning (MSAW)—A function of the ARTS III computer that aids the controller by alerting him when a tracked Mode C equipped aircraft is below or is predicted by the computer to go below a predetermined minimum safe altitude.

minimum weather condition—(*See* VFR FLIGHT WEATHER MINIMUMS IN CONTROLLED AIRSPACE AND VFR FLIGHT WEATHER MINIMUMS IN UNCONTROLLED AIRSPACE.)

MIRL—Medium intensity runway lights.

MLS—Microwave Landing System.

MM—Middle marker ILS.

MOCA—Minimum obstruction clearance altitude.

mode—The letter or number assigned to a specific pulse spacing of radio signals transmitted or received by ground interrogator or airborne transponder components of the Air Traffic Control Radar Beacon System (ATCRBS). Mode A (military Mode 3) and Mode C (altitude reporting) are used in air traffic control.

monoplane—An airplane having one supporting surface.

Morse code—(*See* RADIOTELEGRAPH (MORSE) CODE, Figure R-3.)

mountain flying—Your first experience of flying over mountainous terrain (particularly if most of your flight time has been over the flatlands of the midwest) could be a never-to-be-forgotten nightmare if proper planning is not done and if you are not aware of the potential hazards awaiting. Those familiar section lines are not present in the mountains; those flat, level fields for forced landings are practically nonexistent; abrupt changes in wind direction and velocity occur; severe updrafts and downdrafts are common, particularly near or above abrupt changes of terrain such as cliffs or rugged areas; even

mountain flying *continued*

the clouds look different and can build up with startling rapidity. Mountain flying need not be hazardous if you follow the recommendations below:

1. File a flight plan. Plan your route to avoid topography which would prevent a safe forced landing. The route should be over populated areas and well-known mountain passes. Sufficient altitude should be maintained to permit gliding to a safe landing in the event of engine failure.

2. Don't fly a light aircraft when the winds aloft, at your proposed altitude, exceed 35 miles per hour. Expect the winds to be of much greater velocity over mountain passes than reported a few miles away from them. Approach mountain passes with as much altitude as possible. Downdrafts of from 1,500 to 2,000 feet per minute are not uncommon on the leeward side.

3. Don't fly near or above abrupt changes in terrain. Severe turbulence can be expected, especially in high wind conditions.

4. Some canyons run into a dead end. Don't fly so far up a canyon that you get trapped. ALWAYS BE ABLE TO MAKE A 180° TURN!

5. Plan your trip for the early morning hours. As a rule, the turbulence begins at about 10 A.M. and grows steadily worse until around 4 P.M., then gradually improves until dark. Mountain flying at night in a single engine light aircraft is asking for trouble.

6. When landing at a high altitude field, the same indicated airspeed should be used as at low elevation fields. Remember that due to the less dense air at altitude, this same indicated airspeed actually results in a higher true airspeed, a faster landing speed, and more important, a longer landing distance. During gusty wind conditions which often prevail at high altitude fields, a power approach and power landing is recommended. Additionally, due to the faster groundspeed, your take-off distance will increase considerably over that required at low altitudes.

mountain reporting service—(*See* LAKE/ISLAND, SWAMP, AND MOUNTAIN REPORTING SERVICE.)

mountain wave—Many pilots go all their lives without understanding what a mountain wave is. Quite a few have lost their lives because of this lack of understanding. One need not be a licensed meteorologist to understand the mountain wave phenomenon.

1. Mountain waves occur when air is being blown over a mountain range or even the ridge of a sharp bluff area. As the air hits the upwind side of the range, it starts to climb, thus creating what is generally a smooth updraft which turns into a turbulent downdraft as the air passes the crest of the ridge. From this point, for many miles downwind, there will be a series of downdrafts and updrafts. Satellite photos of the Rockies have shown mountain waves exending as far as 700 miles downwind of the range. Along the east coast area, such photos of the Appalachian chain have picked up the mountain wave phenomenon over a hundred miles eastward. All it takes to form a mountain wave is wind blowing across the range at 15 knots or better at an intersection angle of not less than 30 degrees.

2. Pilots from flatland areas should understand a few things about mountain waves in order to stay out of trouble. When approaching a mountain range from the upwind side (generally the west), there will usually be a smooth updraft; therefore, it is not quite as dangerous an area as the lee of the range. From the leeward side, it is always a good idea to add an extra thousand feet or so of altitude because downdrafts can exceed the climb capability of the aircraft. Never expect an updraft when approaching a mountain chain from the leeward. Always be prepared to cope with a downdraft and turbulence.

3. When approaching a mountain ridge from the downwind side, it is recommended that the ridge be approached at approximately a 45 degree angle to the horizontal direction of the ridge. This permits a safer retreat from the ridge with less stress on the aircraft should severe turbulence and downdraft be experienced. If severe turbulence is encountered, simultaneously reduce power and adjust pitch until aircraft approaches maneuvering speed, then adjust power and trim to maintain maneuvering speed and fly away from the turbulent area.

movement area—The runways, taxiways, and other areas of an airport which are used for taxiing, takeoff, and landing of aircraft, exclusive of loading ramps and parking areas. At those airports with a tower, specific approval for entry onto the movement area must be obtained from ATC.

MRA—Minimum reception altitude.

mrtm—Maritime.

MSL—Mean sea level.

multicom—A mobile service used to provide communications essential to conduct of activities being performed by or directed from private aircraft. Where there is no tower, FSS or UNICOM station on the airport, use MULTICOM frequency 122.9 for self-announce procedures. Such airports will be identified in appropriate aeronautical information publications.

muni—Municipal.

MVFR—Marginal VFR: ceiling 1,000 to 3,000 feet and/or visibility 3 to 5 miles inclusive.

mxd—Mixed.

N

N in the phonetic alphabet is November (no-vem-ber).

N—Preceding numbers of an aircraft represents an aircraft of United States registry.

nacelle—Enclosed shelter for a powerplant or personnel. Usually secondary to the fuselage or cabin.

national airspace system (NAS)—The common system of air navigation and air traffic control, encompassing communications facilities, air navigation facilities, airways, controlled airspace, special use airspace, and flight procedures, and authorized by Federal Aviation Regulations for domestic and international aviation.

National Flight Data Center (NFDC)—A facility in Washington D.C., established by FAA to operate a central aeronautical information service for the collection, validation, and dissemination of aeronautical data in support of the activities of government, industry, and the aviation community. The information is published in the *National Flight Data Digest.* (*See* NATIONAL FLIGHT DATA DIGEST.)

National Flight Data Digest (NFDD)—A daily (except weekends and federal holidays) publication of flight information appropriate to aeronautical charts, aeronautical publications, Notices to Airmen, or other media serving the purpose of providing operational flight data essential to safe and efficient aircraft operations.

national oceanic and atmospheric administration—An agency within the United States Department of Commerce incorporating the Weather Bureau, Coast and Geodetic Survey, and many other environmental science services.

national search and rescue plan—An interagency agreement whose purpose is to provide for the effective utilization of all available facilities in all types of search and rescue missions.

nautical mile—1,852 meters or 1.15 statute miles.

NAVAID—(*See* NAVIGATIONAL AID.)

NAVAID, classes—VOR, DME, and TACAN aids are classed according to their operational use. There are three classes: T (terminal), L (low altitude), and H (high altitude). The normal service range for the T, L, and H class aids is included in the table under NAVAID, RANGES. Certain operational requirements make it necessary to use some of these aids at greater service ranges than are listed in the table. Extended range is made possible through flight inspection determinations. Some aids also have lesser service range due to location, terrain, frequency protection, etc. Restrictions to service range are listed in the *Airport/Facility Directory.*

NAVAID, ranges—

VOR/DME/TACAN NAVAIDS

NORMAL USABLE ALTITUDES AND RADIUS DISTANCES

Class	Altitudes	Distance (miles)
T	12,000′ and below	25
L	below 18,000′	40
H	below 18,000′	40
H	18,000′—FL 450	130
H	FL 450—FL 600	100

NAVAID, ranges *continued*

NONDIRECTIONAL RADIO BEACON (NDB)
USABLE RADIUS DISTANCES FOR ALL ALTITUDES

Class	Distance (miles)
compass locator	15
MH	25
H	50*
HH	75

*Service range of individual facilities may be less than 50 miles. See restrictions to service range in the *Airport/Facility Directory.*

navigable airspace—Airspace at and above minimum flight altitudes, including airspace needed for safe takeoffs and landings.

navigation—Required knowledge of map reading, pilotage, and radio aids to VFR flight are part of the questions covered by the private pilot examination.

navigational aid (NAVAID)—Any visual or electronic device airborne or on the surface which provides point to point guidance information or position data to aircraft in flight.

NDB—Nondirectional radio beacon.

near midair collision reporting—The Federal Aviation Administration is continuing to encourage the transmission of near midair collision reports; however, enforcement of applicable Federal Aviation Administration regulations that might be identified during the investigation of such incidents will be pursued. The agency is vitally interested in all near midair collision incidents. Each reported incident is thoroughly investigated by the agency as soon as received in accordance with the established procedures. In order to ensure expeditious handling, all airmen are urged to report each incident immediately to: (1) nearest Federal Aviation Administration air traffic control facility or flight service station by radio; (2) telephone report at next point of landing to nearest Federal Aviation Administration air traffic control facility or flight service station; (3) written in lieu of (1) and (2) above to the nearest air carrier district office or general aviation district office.

negative—Controller's radio phraseology meaning "No," or "permission not granted," or "that is not correct."

negative contact—Used by pilots to inform ATC that:

1. Previously issued traffic is not in sight. It may be followed by the pilot's request for the controller to provide assistance in avoiding the traffic.

2. They were unable to contact ATC on a particular frequency.

NFCT—Nonfederal control tower. Use same procedures as at tower-controlled airports. (*See* TOWER-CONTROLLED AIRPORTS.)

ngt—Night.

night—The time between the end of evening civil twilight and the beginning of morning civil twilight as published in the *American Air Almanac* and converted to local time.

night flight experience—Ten takeoffs and landings to a full stop and three hours of dual night instruction is required of pilots desiring a night flying endorsement on their certificates. (*See* PRIVATE PILOT, PREREQUISITES FOR CERTIFICATE.)

NMI—Nautical mile(s).

nmrs—Numerous.

no gyro approach/vector—A radar approach/vector provided in case of a malfunctioning gyrocompass or directional gyro. Instead of providing the pilot with headings to be flown, the controller observes the radar track and issues control instructions "turn right/left" or "stop turn," as appropriate.

nonapproach control tower—Authorizes aircraft to land or take off at the airport controlled by the tower, or to transit the airport traffic area. The primary function of a nonapproach control tower is the sequencing of aircraft in the traffic pattern and on the landing area. Nonapproach control towers also separate aircraft operating under instrument flight rules clearances from approach controls and centers. They provide ground control services to aircraft, vehicles, personnel, and equipment on the airport movement area.

nondirectional radio beacon (NDB)—

1. A low- or medium-frequency radio beacon transmits nondirectional signals whereby the pilot of an aircraft properly equipped can

nondirectional radio beacon (NDB) ***continued***

determine his bearing and "home" on the station. These facilities normally operate in the frequency band of 190 to 535 kHz and transmit a continuous carrier with either 400 Hz or 1020 Hz modulation keyed to provide identification except during voice transmission.

2. When a radio beacon is used in conjunction with the instrument landing system markers, it is called a compass locator.

3. All radio beacons except the compass locators transmit a continuous three-letter identification in code except during voice transmissions. Compass locators transmit a continuous two-letter identification in code. The first and second letters of the three-letter location identifier are assigned to the front course outer marker compass locator (LOM), and the second and third letters are assigned to the front course middle marker compass locator (LMM). Example: ATLANTA, ATL, LOM-AT, LMM-TL.

4. Voice transmissions are made on radio beacons unless the letter "W" (without voice) is included in the class designator (HW).

5. Radio beacons are subject to disturbances that may result in erroneous bearing information. Such disturbances result from intermittent/unpredictable signal propagation due to such factors as lightning, precipitation static, etc. At night radio beacons are vulnerable to interference from distant stations. Nearly all disturbances which affect the ADF bearing also affect the facility's identification: noisy identification usually occurs when the ADF needle is erratic; voice, music, or erroneous identification will usually be heard when a steady false bearing is being displayed. Since ADF receivers do not have a "FLAG" to warn the pilot when erroneous bearing information is being displayed, the pilot should continuously monitor the NDB's identification.

nonprecision approach procedure—Standard instrument approach procedure in which no electronic glide slope is provided.

nonradar—Precedes other terms and generally means without the use of radar.

nontower airports—

1. Preparatory to landing at an airport without an operating control tower, but at which either an FSS or a UNICOM is located, pilots

should contact the FSS or UNICOM for traffic advisories, wind, runway in use, and traffic flow information. CAUTION: all aircraft may not be communicating with the FSS or UNICOM. They can only issue traffic advisories on those they are aware of.

2. At those airports not having a tower, FSS or UNICOM, visual indicators, if installed, provide the pilot with landing information. (*See* SEGEMENTED CIRCLE, WIND DIRECTION INDICATORS, LANDING DIRECTION INDICATORS, LANDING STRIP INDICATORS, TRAFFIC PATTERN INDICATORS, RIGHT-OF-WAY APPROACHING AIRPORT.)

nontower airports, airport advisories at—There is no substitute for alertness or, in controller jargon, "Keeping Your Head on a Swivel," while in the vicinity of an airport. This visual alertness can be supplemented with audio alertness by tuning to the appropriate radio frequency and adhering to the following suggested procedures:

An airport may have a full- or part-time tower or flight service station (FSS) located on the airport or a full- or part-time UNICOM station or no aeronautical station. There are three ways for a pilot to communicate his intentions or obtain airport/traffic information when operating at an airport that does not have an operating tower—by communicating with an FSS, a UNICOM operator, or by making a self-announce broadcast.

The key to communicating at an uncontrolled airport is selection of the correct common frequency. The contraction *CTAF,* which stands for Common Traffic Advisory Frequency, is synonymous with this program. A CTAF is a frequency designated for the purpose of carrying out airport advisory practices while operating to or from an uncontrolled airport. The CTAF may be a UNICOM, MULTICOM, FSS, or tower frequency and is identified in appropriate aeronautical publications.

The CTAF frequency for a particular airport can be obtained by contacting any FSS.

Airport Advisory Service (AAS) is a service provided by an FSS physically located on an airport which does not have a control tower or where the tower is temporarily closed or operated on a part-time basis. The CTAF for FSSs which provide this service will be disseminated in appropriate aeronautical publications.

In communicating with a CTAF FSS, establish two-way communications before transmitting outbound/inbound intentions or information.

nontower airports, airport advisories at *continued*

Departing aircraft should state the aircraft type, full identification number, type of flight planned, i.e., VFR or IFR and the planned destination or direction of flight. Report before taxiing and before taking runway for takeoff. If communications with a UNICOM are necessary after initial report to FSS, return to FSS frequency for traffic update.

A CTAF FSS provides wind direction and velocity, favored or designated runway, altimeter setting, known traffic, notices to airmen, airport taxi routes, airport traffic pattern information, and instrument approach procedures. These elements are varied so as to best serve the current traffic situation. Some airport managers have specified that under certain wind or other conditions designated runways be used. Pilots using other than the favored or designated runway should advise the FSS immediately.

CAUTION: All aircraft in the vicinity of an airport may not be in communication with the FSS.

Self-announce is a procedure whereby pilots broadcast their position or intended flight activity or ground operation in the blind on the designated CTAF. This procedure is used primarily at airports which do not have an FSS or UNICOM station on the airport. The self-announce procedure should also be used if a pilot is unable to communicate with the designated CTAF FSS or UNICOM. It should be noted that aircraft operating to or from another nearby airport may be making self-announce broadcasts on the same UNICOM or MULTICOM frequency. To help identify one airport from another, the airport name should be spoken at the beginning and end of each self-announce transmission.

If an airport has a tower and it is temporarily closed, or operated on a part-time basis and there is no FSS on the airport or the FSS is closed, use the CTAF (usually the tower local control frequency) to self-announce your position or intentions. *If there is a UNICOM station in operation on the airport, the wind direction and runway in use should be obtained from the UNICOM station. Then return to and monitor the CTAF and make self-announce broadcasts as appropriate.*

(NOTE: The wind direction and runway information may not be available on UNICOM frequency 122.950).

Recommended phraseologies and communications procedures are as follows: *Aircraft Departing.* When ready to taxi, the pilot should notify the station of his intentions. Except for scheduled air carriers or other frequent users of the airport, this information should include not only the aircraft identification, but also the aircraft type, location, type of flight planned (VFR or IFR), and destination. Example: Aircraft: GRAND FORKS RADIO, COMANCHE NOVEMBER SIX ONE THREE EIGHT, ON TERMINAL BUILDING RAMP, READY TO TAXI, VFR TO DULUTH, OVER. Station: COMANCHE NOVEMBER (Station may or may not repeat this) SIX ONE THREE EIGHT, GRAND FORKS RADIO, ROGER, WIND THREE TWO ZERO DEGREES AT TWO FIVE, FAVORING RUNWAY THREE ONE, ALTIMETER THREE ZERO ZERO ONE, CESSNA ONE-SEVENTY ON DOWNWIND LEG MAKING TOUCH AND GO LANDINGS ON RUNWAY THREE ONE. *OVER. Aircraft Arriving.* When operating VFR, a pilot should transmit position and altitude information to the FSS when 10 miles from the airport. Example: Aircraft: GRAND FORKS RADIO, TRIPACER NOVEMBER ONE SIX EIGHT NINER, OVER KEY WEST, TWO THOUSAND, LANDING GRAND FORKS, OVER. Station: TRIPACER NOVEMBER (Station may or may not repeat this) ONE SIX EIGHT NINER, GRAND FORKS RADIO, OVER KEY WEST AT TWO THOUSAND, WIND ONE FIVE ZERO DEGREES AT FOUR, DESIGNATED RUNWAY FIVE, DC-3 TAKING OFF RUNWAY FIVE, BONANZA ON DOWNWIND LEG RUNWAY FIVE MAKING TOUCH AND GO LANDINGS, COMANCHE DEPARTED RUNWAY ONE SEVEN AT ONE SIX PROCEEDING EAST-BOUND, OVER. (*Also see* MULTICOM, UNICOM.)

NORDO—(*See* LOST COMMUNICATIONS/TWO-WAY RADIO COMMUNICATIONS.)

normal climb—Indicated airspeeds for this type of climb are given in each aircraft's owner's manual.

northerly turn error—When turning to or from headings of north and south, is caused by magnetic dip.

noseheavy—A condition of trim in an airplane in which the nose tends to sink when the elevator control is released.

nose-over—The turning of an airplane on its back on the ground by rolling over the nose.

nosewheel—A swiveling or steerable wheel mounted forward in tricycle-geared airplanes.

notice to airmen (NOTAM)—A notice containing information (not known sufficiently in advance to publicize by other means) concerning the establishment, condition, or change in any component (facility, service, or procedure of, or hazard in the National Airspace System) the timely knowledge of which is essential to personnel concerned with flight operations.

1. NOTAM(D)—A NOTAM given (in addition to local dissemination) distant dissemination via teletypewriter beyond the area of responsibility of the Flight Service Station. These NOTAMS will be stored and repeated hourly until canceled.
2. NOTAM(L)—A NOTAM given local dissemination by voice (teletypewriter where applicable), and a wide variety of means such as: TelAutograph, teleprinter, facsimile reproduction, hot line, telecopier, telegraph, and telephone to satisfy local user requirements.
3. FDC NOTAM—A notice to airmen, regulatory in nature, transmitted by NFDC and given all-circuit dissemination.

notification of arrival—Required by VFR pilots who have filed a flight plan.

notification of serious accidents—Or in-flight hazards of fire or flight control system malfunction, is required to be made immediately to the nearest Federal Aviation Administration Bureau of Safety field office.

Nr—number

NSARP—(*See* NATIONAL SEARCH AND RESCUE PLAN.)

ntc—Notice.

numbers—(*See* FIGURES.)

numerous targets vicinity (location)—A traffic advisory issued by Air Traffic Control to advise pilots that targets on the radar scope are too numerous to issue individually.

O

O in the phonetic alphabet is Oscar (oss-cah).

obsc—Obscure.

obstacle—An existing object, object of natural growth, or terrain at a fixed geographical location, or which may be expected at a fixed location within a prescribed area, with reference to which vertical clearance is or must be provided during flight operation.

obstn—Obstruction.

obstructions—Indicated by:

1. Red flashes during the day and steady red at night meaning the presence of an obstruction or obstructions to air navigation or an area on the ground used for purposes hazardous to air navigation. (Reference Advisory Circular No 70/7460-1F, "Obstruction Marking and Lighting," issued by the Federal Aviation Administration).

2. Steady burning red lights, employed near airports to mark obstructions, and also used to supplement flashing lights in marking en route obstructions.

3. High intensity flashing white lights, being used to identify some supporting structures of overhead transmission lines located across rivers, chasms, gorges, etc. These lights flash in a middle, top, lower light sequence at approximately 60 flashes per minute. The top light is normally installed near the top of the supporting structure, while the lower light indicates the approximate lower portion of the wire

obstructions ***continued***

span. The lights are beamed toward the companion structure and identify the area of the wire span.

4. High intensity flashing white lights, beginning to be employed to identify tall structures, such as chimneys and towers, as obstructions to air navigation. The lights provide a 360 degree coverage about the structure at 40 flashes per minute and consist of from 1 to 7 levels of lights, depending upon the height of the structure. Where more than one level is used the vertical banks flash simultaneously. (*Also see* Figure O-1.)

occluded front—When a warm and a cold front join.

ocln—Occlusion.

ocnl—Occasional.

ocr—Occur.

oct—Octane.

off-route vector—A vector by ATC which takes an aircraft off a previously assigned route. Altitudes assigned by ATC during such vectors provide required obstacle clearance.

offset parallel runways—Staggered runways having centerlines which are parallel.

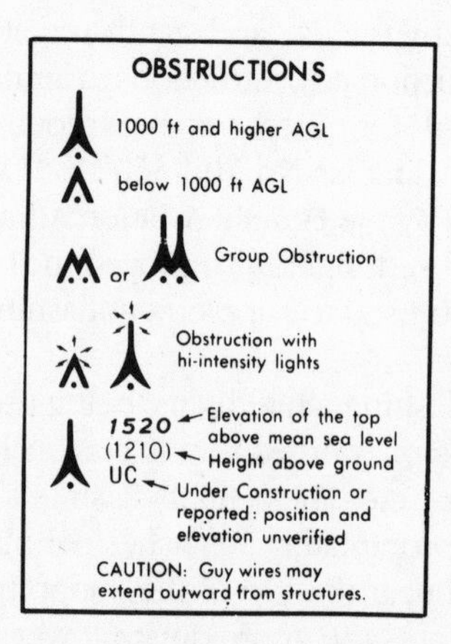

Figure O-1. Obstructions identifying tall structures.

OM—Outer marker instrument landing system (ILS).

omni—Abbreviation referring to VHF omnidirectional range (VOR) since many pilots refer to VORs as omni stations.

on-course indication—An indication on an instrument which provides the pilot a visual means of determining that the aircraft is located on the centerline of a given navigational track, or an indication on a radar scope that an aircraft is on a given track.

oper—Operate.

operation of aircraft rotating beacons—There have been several incidents in which small aircraft were overturned or damaged by prop/jet blast forces from taxiing large aircraft. A small aircraft taxiing behind any large aircraft with its engines operating could meet with the same results. In the interest of preventing ground upsets and injuries to ground personnel due to prop/jet engine blast forces, the Federal Aviation Administration has recommended to air carriers/ commercial operators that they establish procedures for the operation of the aircraft rotating beacon any time the engines are in operation. General aviation pilots utilizing aircraft equipped with rotating beacons are also encouraged to participate in this program and operate the beacon any time the aircraft engines are in operation as an alert to other aircraft and ground personnel that prop/jet engine blast forces may be present. Caution must be exercised by all personnel not to rely solely on the rotating beacon as an indication that aircraft engines are in operation, since participation in this program is voluntary.

operation raincheck—Pilots are encouraged to visit air traffic facilities (Towers, Centers, and FSSs) and participate in Operation Raincheck. Operation Raincheck is a program designed to familiarize pilots with the ATC system, its functions, responsibilities and benefits. On rare occasions, facilities may not be able to approve a visit because of ATC workload or other reasons. It is therefore requested that pilots contact the facility prior to the visit and advise of the number of persons in the group, the time and date of the proposed visit and the primary interest of the group. With this information available, the facility can prepare an itinerary and have someone available to guide the group through the facility.

operations—At airports with control towers, operations require that aircraft be equipped with two-way radio. If, however, radio fails in flight and weather conditions are at or above the basic VFR minimums, pilot may maintain visual contact with the tower and receive light gun clearance to land.

opn—Operation.

ops—Operations.

option approach—An approach requested and conducted by a pilot which will result in either a touch-and-go, missed approach, low approach, stop-and-go, or full stop landing.

orientation—The act of fixing position or attitude by visual or other reference.

oscillation error—Caused by the erratic swinging of the compass card, resulting from either rough air or poor pilot technique. When reading compass, be sure aircraft is as steady as possible.

otlk—Outlook.

otrw—Otherwise.

out—Pilot or controller expression meaning the conversation is ended and no response is expected.

outer fix—A fix in the destination terminal area, other than the approach fix, to which aircraft are normally cleared by an air route traffic control center or a terminal area traffic control facility, and from which aircraft are cleared to the approach fix or final approach course.

outer marker (OM)—A marker beacon at or near the glide slope intercept altitude of an ILS approach. It is keyed to transmit two dashes per second on a 400 Hz tone which is received aurally and visually by compatible airborne equipment. The OM is normally located 4 to 7 miles from the runway threshold on the extended centerline of the runway.

ovc—Overcast.

over—Pilot or controller expression meaning transmission is ended; I expect a response.

overshoot—To fly beyond a designated area or mark.

overtaking—Each aircraft that is overtaken has the right-of-way, and each pilot of an overtaking aircraft shall alter course to the right and pass well clear.

over the top—Above the layer of clouds forming the ceiling.

ovrn—Overrun.

owner's flight manual—(*See* AIRPLANE OWNER'S FLIGHT MANUAL.)

oxygen—(*See* HYPOXIA.)

P

P in the phonetic alphabet is Papa (pah-pah).

P—Symbol for polar air mass.

PAN—The international radio-telephone urgency signal. When repeated three times indicates uncertainty or alert, followed by nature of urgency. (*See* MAYDAY.)

panic—Leads to unwise and precipitous action. Panic can be controlled. If lost or in some other predicament, call the emergency frequency 121.5 and ask for help.

PAPI—Precision approach path indicator. (*See* VISUAL APPROACH SLOPE SYSTEMS.)

PAR—Precision approach radar.

parallel runways—Two or more runways at the same airport whose centerlines are parallel. In addition to runway number, parallel runways are designated as L (left) and R (right) or, if three parallel runways exist, L (left), C (center), and R (right).

passengers, carrying of—No person may act as pilot carrying passengers unless he has, within the preceding 90 days, made at least 3 takeoffs and landings to a full stop in the aircraft of the same type, class, and category. No private pilot may carry passengers for hire, but he may share operating expenses of a flight with his passengers. No student pilot acting as pilot in command may carry *any* passengers.

pattern direction—At all airports, is in a rectangular pattern from the left unless special obstructions or other considerations require

pattern direction *continued*

approach from the right. This fact will be indicated by the markings in segmented circle on the airport and/or a flashing amber light signal.

PATWAS—(*See* PILOTS AUTOMATIC WEATHER ANSWERING SERVICE.)

pbl—Probable.

PCA—Positive control area.

pcpn—Precipitation.

periodic inspection—Airframe and engine inspection is required every 12 calendar months or every 100 hours on aircraft for hire. (*Also see* PROGRESSIVE INSPECTION.)

permly—Permanently.

personnel locator beacon—Small, portable device which can be carried on the person or as a part of the personal equipment of the flight crew or occupants in case of the aircraft's being downed in an emergency.

P-factor—Asymmetrical thrust of the propeller when aircraft is in a climbing attitude, which has a tendency to pull the aircraft to the left. (*Also see* TORQUE.)

phonetic alphabet—(*See* ALPHABET, PHONETIC.)

physical standards—Require that a student or private pilot hold an active third class medical certificate.

PIC—Pilot in command.

pilot—One who operates the controls of an airplane in flight.

pilot, age minimum for—Private pilot, 17; student, 16.

pilot authority—The pilot in command of an aircraft is directly responsible for, and the final authority on, the operation of that aircraft.

pilot certificate—An applicant who meets the requirements of the Federal Aviation Regulations is issued a pilot certificate with appropriate aircraft ratings. (*See also* STUDENT PILOT, DURATION OF CERTIFICATE.)

Pilot certificates, holders of, change of address regulations—All holders of pilot certificates are required within 30 days of a permanent home address change to notify, in writing, the Federal Aviation Administration Airman Certification Branch, Box 25082, Oklahoma City, OK 73125.

pilot control of airport lighting—(*See* AIRPORT LIGHTING, PILOT CONTROL OF.)

pilot logbooks—Must maintain a reliable record of all flight experience required for pilot certificates or ratings.

pilot maneuvers in traffic pattern—On occasion it may be necessary for a pilot to maneuver his aircraft to maintain spacing with the traffic he has been sequenced to follow. The controller can anticipate minor maneuvering such as shallow "S" turns. The controller cannot, however, anticipate a major maneuver such as a 360 degree turn. If a pilot makes a 360 degree turn after he has obtained a landing sequence, the result is usually a gap in the landing interval and more importantly it causes a chain reaction which may result in a conflict with following traffic and interruption of the sequence established by the tower or approach controller. Should a pilot decide he needs to make maneuvering turns to maintain spacing behind a preceding aircraft, he should always advise the controller if at all possible. Except when requested by the controller or in emergency situations, a 360 degree turn should never be executed in the traffic pattern or when receiving radar service without first advising the controller.

pilot responsibility, vortex—Government and industry groups are making concerted efforts to minimize the hazards of trailing vortices. However, the flight disciplines necessary to assure vortex avoidance during VFR operations must be exercised by the pilot. Vortex visualization and avoidance is equal in importance to traffic avoidance. (*See* VORTEX AVOIDANCE PROCEDURES.)

pilot visits—Pilots are encouraged to visit air traffic facilities—towers, centers, and flight service stations. On rare occasions, facilities may not be able to approve a visit because of workload or other reasons. It is therefore requested that pilots contact the facility prior to the visit, giving the number of persons in the group, the time and date of the proposed visit, and the primary interest of the group. With this information available, the facility can prepare an itinerary and have someone available to guide the group through.

pilot weather reports (PIREPS)—

1. Whenever ceilings are at or below 5,000 feet, visibilities at or below 5 miles, thunderstorms, icing, turbulence, or wind shear are reported or forecast, Federal Aviation Administration stations are required to solicit and collect PIREPS which describe conditions aloft. Pilots are urged to cooperate and volunteer reports of wind shear, cloud tops, upper cloud layers, thunderstorms, ice, turbulence, strong winds, and other significant flight condition information. Such conditions observed between weather reporting stations are vitally needed. The PIREPS should be given to the Federal Aviation Administration ground facility with which communication is established. i.e., EFAS, FSS, ARTCC, or terminal ATC. In addition to complete PIREPS, pilots can materially help round out the in-flight weather picture by adding to routine position reports, both VFR and IFR, the type of aircraft, and the following phrases as appropriate: ON TOP; BELOW OVERCAST; WEATHER CLEAR; LIGHT, MODERATE (or HEAVY) ICING; LIGHT, MODERATE, SEVERE, EXTREME TURBULENCE; FREEZING RAIN (or DRIZZLE); THUNDERSTORM (location); BETWEEN LAYERS; ON INSTRUMENTS; ON AND OFF INSTRUMENTS.

2. If pilots are not able to make PIREPS by radio, reporting upon landing of the in-flight conditions encountered to the nearest flight service station or weather service station will be helpful.

pilots automatic telephone weather answering service (PATWAS)—

1. At some locations the numbers of pilots requiring light weather briefings are too numerous for person-to-person briefings. To assist in this important service, recorded weather briefings are available at several locations. This service is called PATWAS.

2. PATWAS locations are found in the Airport/Facility Directory under the FSS and National Weather Service Telephone Numbers section.

pilot's discretion—When used by a controller in conjunction with altitude assignments, means that ATC has offered the pilot the option of starting climb or descent whenever he wishes and conducting the climb or descent at any rate he wishes. He may temporarily level off at any intermediate altitude. However, once he has vacated an altitude he may not return to that altitude.

pilot's personal checklist—Aircraft accident statistics show that pilots should be conducting preflight checklists on themselves as well as their aircraft, for pilot impairment contributes to many more accidents than failures of aircraft systems. A personal checklist that can be easily committed to memory, which includes all of the categories of pilot impairment as discussed in this section, is being distributed by the FAA in the form of a wallet-sized card:

PERSONAL CHECKLIST

I'm physically and mentally safe to fly—not being impaired by:

Illness,
Medication,
Stress,
Alcohol,
Fatigue,
Emotion.

pilotage—Navigation by visual reference to landmarks.

PIREPS—(*See* PILOT WEATHER REPORTS.)

pitch (airplane)—Angular displacement of the longitudinal axis with respect to the horizon.

pitch (propeller)—The angle of its blades measured from its plane of rotation.

pitot tube—A tube exposed to the air stream for measuring impact pressure.

plane—An airfoil section for deflection of air; surface or field of action in any two dimensions only; to move over the water (seaplanes) so the weight is supported by dynamic reation of the water, rather than by displacement.

plasi—Pulsed light approach slope indicator. (*See* VISUAL APPROACH SLOPE SYSTEMS.)

p-line—Pole line; also power line.

porpoising—Pitching while planing

position lights—Required on all aircraft operated at night.

positive control area—Airspace so designated as positive control area in FAR Part 71. This area includes specified airspace within the conterminous U.S. from 18,000 feet to and including FL 600, excluding Santa Barbara Island, Farallon Island, and that portion south of latitude 25 degrees 04 minutes north. In Alaska, it includes the airspace over the State of Alaska from 18,000 feet to and including FL 600, but not including the airspace less than 1,500 feet above the surface of the earth and the Alaskan Peninsula west of longitude 160 degrees 00 minutes west. Rules for operating in Positive Control Area are found in Federal Aviation Regulations Part 91.

power approach—Landing, using throttle rather than glide.

PPR—Prior permission required.

practice instrument approach procedures—Pilot request to practice instrument approach procedures may be approved by ATC subject to traffic and workload conditions.

precipitation—Any or all forms of water particles (rain, sleet, hail, or snow) that fall from the atmosphere and reach the surface.

precision approach—A descent in an approved procedure where the navigation facility alignment is normally on the runway centerline, and glide slope information is provided such as ILS or PAR.

precision landing—(*See* ACCURACY LANDINGS.)

precision radar—

1. Precision approach radar is designed to be used as a landing aid, rather than an aid for sequencing and spacing aircraft. PAR equipment may be used as a primary landing aid, or it may be used to monitor other types of approaches. It is designed to display range, azimuth, and elevation information.

2. Two antennae are used in the PAR array, one scanning a vertical plane, and the other scanning horizontally. Since the range is limited to 10 miles, azimuth to 20 degrees, and elevation to 7 degrees, only the final approach area is covered. Each scope is divided into two parts. The upper half presents altitude and distance information, and the lower half presents azimuth and distance.

preflight briefing—Required of every pilot to familiarize himself with all available information concerning that flight.

preflight check—Each aircraft must be checked before takeoff. Includes visual inspection of the wings, tail, propeller, landing gear, fuel tanks, and oil level, pitot tube, stall warning, and fuel tank vents for stoppages, gas line drain and light operations. If night flight is planned, be sure a flashlight is available.

pressure altitude—(*See* ALTITUDE.)

private pilot aeronautical experience—(*See* PRIVATE PILOT, PREREQUISITES FOR CERTIFICATE.)

private pilot certificate, duration—Currently has no expiration date but requires a biennial flight review to maintain. (*See* BIENNIAL FLIGHT REVIEW.)

private pilot certificate, written test—An applicant for a private pilot certificate must, within 24 months prior to his flight test, pass a written test on Federal Aviation Regulations on general operating and air traffic rules and accident reporting rules of the National Transportation Safety Board; practical aspects of cross-country flying including navigation, radio communication, and emergency procedures; recognizing dangerous weather conditions and evaluating weather reports; and general safety practices.

private pilot eligibility—(*See* PRIVATE PILOT, PREREQUISITES FOR CERTIFICATE.)

private pilot, limitations on, for compensation or hire—(*See* PASSENGERS, CARRYING OF.)

private pilot medical certificate, duration—The private or student pilot is required to have a third class medical certificate which expires at the end of the last day of the 24th month after the month in which it is issued. (*See* MEDICAL CERTIFICATES.)

private pilot, prerequisites for certificate—17 years of age, ability to read, speak, and understand English language, a third-class medical certificate; must have passed a written and flight test, and have a total of 40 hours flight, instruction, and solo time. This time must include the following:

1. Twenty hours of flight instruction from an authorized flight instructor, including at least—

   a. Three hours of cross-country.

private pilot, prerequisites for certificate ***continued***

b. Three hours at night, including 10 takeoffs and landings for applicants seeking night flying privileges; and

c. Three hours in airplanes in preparation for the private pilot flight test within 60 days prior to that test. An applicant who does not meet the night flying requirement in paragraph 1b of this section is issued a private pilot certificate bearing the limitation "Night flying prohibited." This limitation may be removed if the holder of the certificate shows that he has met the requirements of paragraph 1b of this section.

2. Twenty hours of solo flight time, including at least—

a. Ten hours in airplanes.

b. Ten hours of cross-country flights, each flight with a landing at a point more than 50 nautical miles from the original departure point. One flight must be of at least 300 nautical miles with landings at a minimum of three points, one of which is at least 100 nautical miles from the original departure point.

c. Three solo takeoffs and landings to a full stop at an airport with an operating control tower.

private pilot certificate, retesting for—Applicant may apply for retesting on presentation of a recommendation form from a certified flight instructor that he has been given additional instruction and is now ready for retesting after first failure. On second failure, applicant must wait 30 days.

procedure words and phrases—(*See* RADIO PROCEDURE WORDS AND PHRASES.)

proficiency test—(*See* BIENNIEL FLIGHT REVIEW and VFR FLIGHT PROFICIENCY.)

progressive inspection—Over a 12-month period may be allowd on approval from the nearest flight standards district office.

prohibited area—Airspace of defined dimensions identified on charts by an area on the surface of the earth within which flight is prohibited.

propeller—Any device for producing thrust in any fluid.

protractor—A device for measuring angles. Generally used in navigation to determine compass courses on a chart.

prst—Persist.

psbl—Possible.

psg—Passing.

ptly—Partly.

pusher—An airplane in which the propeller is mounted aft of the engine and pushes the air away from it.

pvl—Prevail.

pylon—A prominent mark, or point, on the ground used as a fix in precision maneuvers.

Q

Q in the phonetic alphabet is Quebec (keh-beck)

quad—Quadrant.

quadrant—A quarter part of a circle, centered on NAVAID, oriented clockwise from magnetic north as follows: NE quadrant 000–089, SE quadrant 090–179, SW quadrant 180–269, NW quadrant 270–359.

R

R in the phonetic alphabet is Romeo (row-me-oh).

rad—Radial.

RADAR (radio detecting and ranging)—A device which, by measuring the time interval between transmission and reception of radio

RADAR (radio detecting and ranging) ***continued***

pulses and correlating the angular orientation of the radiated antenna beam or beams in azimuth and/or elevation, provides information on range, azimuth and/or elevation of objects in the path of the transmitted pulses.

1. Primary Radar—A radar system in which a minute portion of a ratio pulse transmitted from a site is reflected by an object and then received back at that site for processing and display at an air traffic control facility.

2. Secondary Radar/Radar Beacon/ATCRBS—A radar system in which the object to be detected is fitted with cooperative equipment in the form of a radio receiver/transmitter (transponder). Radar pulses transmitted from the searching transmitter/receiver (interrogator) site are received in the cooperative equipment and used to trigger a distinctive transmission from the transponder. This reply transmission, rather than a reflected signal, is then received back at the transmitter/receiver site for processing and display at an air traffic control facility. (*See* TRANSPONDERS.)

radar advisory—indicates that the provision of advice and information is based on radar observation.

radar assistance to VFR aircraft—Radar equipped Federal Aviation Administration air traffic control facilities provide radar assistance and navigation service (vectors) to VFR aircraft provided the aircraft can communicate with the facility, is within radar coverage and can be radar identified. Radar covers most areas of the U.S. above 4,000–4,500 feet.

radar beacon phraseology—(*See* TRANSPONDER OPERATION.)

radar beacon (secondary radar)—(*See* RADAR.)

radar beacon system—(*See* AIR TRAFFIC CONTROL RADAR BEACON SYSTEM.)

radar contact—Used by air traffic controllers, indicates that an aircraft is identified on the radar display and that radar service can be provided until radar identification is lost or radar service is terminated. When the aircraft is informed of "radar contact" it automatically discontinues reporting over compulsory reporting points.

radar contact lost—Used by ATC to inform a pilot that radar identification of his aircraft has been lost. The loss may be attributed to several things including the aircraft merging with weather or ground clutter, the aircraft flying below radar line of sight, the aircraft entering an area of poor radar return, or a failure of the aircraft transponder or ground radar equipment.

radar flight following—The radar tracking of identified aircraft targets and the observation of the progress of such flights sufficiently to retain identity.

radar handoff—That action whereby radar identification of, radio communications with, and, unless otherwise specified, control responsibility for an aircraft is transferred from one controller to another without interruption of radar flight following.

radar identification—The process of ascertaining that a radar target is the radar return from a particular aircraft.

radar, precision—(*See* PRECISION RADAR.)

radar programs—For VFR aircraft at terminals. (*See* TERMINAL RADAR PROGRAMS FOR VFR AIRCRAFT.)

radar service—A term which encompasses one or more of the following services, based on the use of radar which can be provided by a controller to a pilot of a radar-identified aircraft:

1. Radar separation—Radar spacing of aircraft in accordance with established minima.

2. Radar navigation guidance—Vectoring aircraft to provide course guidance.

3. Radar monitoring—The radar flight following of aircraft, whose primary navigation is being performed by the pilot, to observe and note deviations from its authorized flight path airway, or route. As applied to the monitoring of instrument approaches from the final approach fix to the runway, it also includes the provision of advice on position relative to approach fixes and whenever the aircraft proceeds outside the prescribed safety zones.

radar service for VFR aircraft in difficulty—Radar equipped Federal Aviation Administration air traffic control facilities provide radar assistance and navigation service (vectors) to VFR aircraft in

radar service for VFR aircraft in difficulty *continued*

difficulty, provided the aircraft can communicate with the facility, is within radar coverage, and can be radar identified. Radar service is automatically terminated when an arriving VFR aircraft, receiving radar services is advised to contact the tower.

radar surveillance—The radar observation of a given geographical area for the purpose of performing some radar function. (*See* SURVEILLANCE RADAR.)

radar traffic informaiton service—A service provided by radar air traffic control facilities. Pilots receiving this service are advised of any radar target observed on the radar display which may be in such proximity to the position of their aircraft or its intended route of flight that it warrants their attention. This service is not intended to relieve the pilot of his responsibility for fontinual vigilance to see and avoid other aircraft.

radar vector—A heading issued to an aircraft to provide navigational guidance by radar.

Radar weather echo intensity levels—Existing radar systems cannot detect turbulence. However, there is a direct correlation between the degree of turbulence and other weather features associated with thunderstorms and the radar weather echo intensity. The National Weather Service has categorized six levels of radar weather echo intensity. The following list gives the weather features likely to be associated with these levels during thunderstorm weather situations:

1. Level 1 (WEAK) and Level 2 (MODERATE)—Light to moderate turbulence is possible with lightning.
2. Level 3 (STRONG)—Severe turbulence possible, lightning.
3. Level 4 (VERY STRONG)—Severe turbulence likely, lightning.
4. Level 5 (INTENSE)—Severe turbulence, lightning, organized wind gusts. Hail likely.
5. Level 6 (EXTREME)—Severe turbulence, large hail, lightning, extensive wind gusts and turbulence.

radial—A magnetic bearing extending from an ADF, a VOR, VORTAC, or TACAN navigation facility.

radio—Expression used both for:

1. A device used for communication and

2. Used to refer to a Flight Service Station, e.g., "Seattle Radio" is used to call Seattle FSS.

radio aids to navigation, symbols—(*See* Figure R-1.)

radio compass—(*See* AUTOMATIC DIRECTION FINDER.)

radio contact procedure—

1. Initiate radio communications with a ground facility by using the following format: (a) identification of the unit being called; (b) identification of the aircraft; (c) the type of message to follow, when this will be of assistance; (d) the word "OVER." Example: NEW YORK RADIO, MOONEY NOVEMBER THREE ONE ONE ONE ECHO, OVER.

2. Reply to callup from a ground facility by using the following format: (a) idenification of the unit initiating the callup; (b) identification of the aircraft; (c) the word "OVER." Example: PITTSBURGH TOWER, CESSNA NOVEMBER TWO SIX FOUR FIVE ZULU OVER. NOTE: The word "OVER" may be omitted if the message obviously requires a reply.

3. Use the same format as for initial callup and reply after communication had been established except, after stating your identification, state the message to be sent or acknowledgment of the message received. The acknowledgment is made with the word "ROGER" or

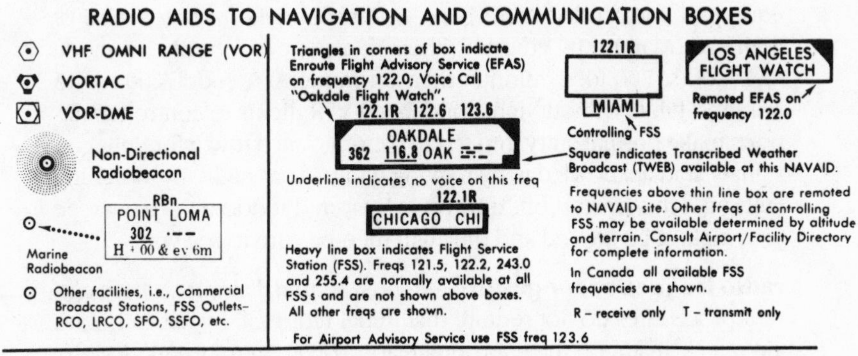

RADIO AIDS TO NAVIGATION AND COMMUNICATION BOXES

VHF OMNI RANGE (VOR)

VORTAC

VOR-DME

Non-Directional Radiobeacon

RBn
POINT LOMA
302
H + 00 & ev 6m

Marine Radiobeacon

Other facilities, i.e., Commercial Broadcast Stations, FSS Outlets–RCO, LRCO, SFO, SSFO, etc.

Triangles in corners of box indicate Enroute Flight Advisory Service (EFAS) on frequency 122.0; Voice Call "Oakdale Flight Watch".

122.1R 122.6 123.6
OAKDALE
362 116.8 OAK

Underline indicates no voice on this freq

122.1R
CHICAGO CHI

Heavy line box indicates Flight Service Station (FSS). Freqs 121.5, 122.2, 243.0 and 255.4 are normally available at all FSSs and are not shown above boxes. All other freqs are shown.

For Airport Advisory Service use FSS freq 123.6

122.1R
MIAMI

Controlling FSS

LOS ANGELES FLIGHT WATCH

Remoted EFAS on frequency 122.0

Square indicates Transcribed Weather Broadcast (TWEB) available at this NAVAID.

Frequencies above thin line box are remoted to NAVAID site. Other freqs at controlling FSS may be available determined by altitude and terrain. Consult Airport/Facility Directory for complete information.

In Canada all available FSS frequencies are shown.

R – receive only T – transmit only

Figure R-1. Radio aids to navigation, symbols.

radio contact procedure *continued*

"WILCO" and pilots are expected to comply with ATC clearances/instructions when they acknowledge by using either "ROGER" or "WILCO." Example: APACHE ONE TWO THREE XRAY, ROGER.

4. After contact has been definitely established, it may be continued without further callup or identification.

5. Pilots operating under provisions of Federal Aviation Regulations, Part 135, ATCO certificate are urged to prefix their aircraft identification with the phonetic word "Tango" on the initial contact with ATC facilities unless they have been assigned Federal Aviation Administration authorized call signs. Example: TANGO AZTEC 2464 ALFA. NOTE: The prefix "Tango" may be dropped on subsequent contacts on the same frequency.

6. Abbreviated call signs may be used ONLY when initiated by the ground station, and will consist of the aircraft type followed by the last three characters of the tail number. Example: "TRIPACER SIX TWO YANKEE."

radio detection and ranging—(*See* RADAR.)

radio failure procedures—Pilots who experience radio communications failure are urged to listen on any operational radio receiver for information broadcast by air traffic control. Controllers have the capability of transmitting on most navigational facilities and do so when an aircraft communications failure is recognized. Pilots of aircraft equipped with coded radar beacon transponders may alert ATC of their radio failure by adjusting their transponder to reply on mode A/3, Code 7700 for 1 minute, then on mode A/3, code 7600 for 15 minutes. Pilots without transponders on VFR flights to controlled airports make normal entry into traffic pattern, watch tower for light gun signals, and acknowledge by rocking wings, or at night, by switching landing lights on and off. If tower is busy and it doesn't seem to see you, make a go-around and they will then be sure to see you.

radio frequency congestion—At busy terminal airports where current procedures do not require that initial radio call by ALL in-bound aircraft be made on the appropriate approach control frequency, the pilot, by adopting the following practice, will decrease frequency congestion and thereby increase safety: monitor the tower local frequency for a sufficient length of time prior to reaching normal callup point to determine wind and runway information. When initial call is made to tower the following information will suffice: identification, position, altitude, and the fact that wind and runway information has been received; the term "have numbers" or a similar phrase should be used by the pilot for this purpose. Additionally, if appropriate VFR reporting point for the runway in use is known, advise tower you will report over that point. Also check the information under the heading automatic terminal information service (ATIS).

radio magnetic indicator (RMI)—An aircraft navigational instrument coupled with a gyro compass or similar compass that indicates the direction of a selected NAVAID and indicates bearing with respect to the heading of the aircraft.

radio procedure words and phrases—The following words and phrases should be used where practicable in radiophone communications:

WORD OR PHRASE	MEANING
ACKNOWLDEGE	"Let me know that you have received and understood this message."
AFFIRMATIVE	"Yes"
CORRECTION	"An error has been made in this transmission. The correct version is . . ."
GO AHEAD	"Proceed with your message."
HOW DO YOU HEAR ME?	Self-explanatory.
I SAY AGAIN	Self-explanatory.
NEGATIVE	"No" or "Permission not granted" or "That is not correct."
OUT	"This conversation is ended and no response is expected."
OVER	"My transmission is ended and I expect a response from you."
READ BACK	"Repeat all of this message back to me."
ROGER	"I have received all of your last transmission." (Acknowledges receipt; shall not be used for other purposes.)
SAY AGAIN	Self-explanatory.
SPEAK LOWER	Self-explanatory.
STAND BY	If used by itself means "I must pause for a few seconds." If the pause is longer than a few seconds, or if "STAND BY" is used to prevent another station from transmitting, it must be followed by the ending "OUT."

radio procedure words and phrases *continued*

THAT IS CORRECT .... Self-explanatory.
VERIFY .... Confirm
WILCO .... I have received your message, understand it, and will comply.
WORDS TWICE .... (a) As a request: "Communication is difficult. Please say every phrase twice."
(b) As information: "Since communication is difficult, every phrase in this message will be spoken twice."

radio receiving and transmitting frequencies—(*See* AIRCRAFT RADIO RECEIVING FREQUENCIES; AIRCRAFT RADIO TRANSMITTING FREQUENCIES.)

radio reception distances—When there are no intervening obstructions to "radio line of sight," an aircraft at 1,000 foot altitude can receive a radio signal approximately 40 nautical miles away; at 3,000 feet 70 nautical miles away; at 5,000 feet (a station) 90 (nautical) miles (away); and at 10,000 feet (a station) 120 (nautical) miles (away). Radar reception distances are usually less; for example, a large city airport can pick up an aircraft at 4,000 feet from 40–50 miles away; although ARSR can cover 200 NMI.

radiotelegraph (Morse) code—(*See* Figure R-2.)

A	•—	Alfa	(AL-FAH)
B	—•••	Bravo	(BRAH-VOH)
C	—•—•	Charlie	(CHAR-LEE) (or SHAR LEE)
D	—••	Delta	(DELL-TAH)
E	•	Echo	(ECK-OH)
F	••—•	Foxtrot	(FOKS-TROT)
G	——•	Golf	(GOLF)
H	••••	Hotel	(HOH-TEL)
I	••	India	(IN-DEE-AH)
J	•———	Juliett	(JEW-LEE-ETT)
K	—•—	Kilo	(KEY-LOH)
L	•—••	Lima	(LEE-MAH)
M	——	Mike	(MIKE)
N	—•	November	(NO-VEM-BER)
O	———	Oscar	(OSS-CAH)
P	•——•	Papa	(PAH-PAH)
Q	——•—	Quebec	(KEH-BECK)
R	•—•	Romeo	(ROW-ME-OH)
S	•••	Sierra	(SEE-AIR-RAH)
T	—	Tango	(TANG-GO)
U	••—	Uniform	(YOU-NEE-FORM) (or OO-NEE-FORM)
V	•••—	Victor	(VIK-TAH)
W	•——	Whiskey	(WISS-KEY)
X	—••—	Xray	(ECKS-RAY)
Y	—•——	Yankee	(YANG-KEY)
Z	——••	Zulu	(ZOO-LOO)
1	•————	Wun	
2	••———	Too	
3	•••——	Tree	
4	••••—	Fow-er	
5	•••••	Fife	
6	—••••	Six	
7	——•••	Sev-en	
8	———••	Ait	
9	————•	Nin-er	
0	—————	Zero	

Figure R-2. Radiotelegraph (Morse) code.

ramp—(*See* APRON/RAMP.)

RAPCON—Radar approach control (United States Air Force).

rarep—Radar report.

RATCC—Radar air traffic control center (Navy).

rate-of-climb indicator—An instrument which indicates the rate of ascent or descent of an airplane.

rating—Statement, part of a certificate setting forth special conditions, privileges, or limitations, such as "single engine land," meaning the pilot is authorized to fly only single engine aircraft and to land on land surfaces.

rating requirements—(*See* PRIVATE PILOT, PREREQUISITES FOR CERTIFICATE.)

RBN—Radio beacon.

RCAG—Remote center air/ground.

RCC—Rescue coordination center.

RCO—Remote communications outlet.

rcv—Receive.

rcvg—Receiving

rcvr—Receiver

rdg—Ridge.

rdo—Radio.

read back—Controller's expression meaning repeat my message back to me.

recent flight experience—(*See* PASSENGERS, CARRYING OF.)

reconst—Reconstruction.

red tinted bands—Usually referred to as magenta; on aeronautical charts, indicate controlled airspace 700 feet above surface.

REIL—Runway end identifier lights.

relctd—Relocated.

release time—a departure time restriction issued to a pilot by ATC when necessary to separate a departing aircraft from the other traffic.

remote communications outlet (RCO)—An unmanned satellite air-to-ground communications station remotely controlled, providing UHF and VHF transmit and receive capability to extend the service range of the FSS.

report—Controller expression used to instruct pilots to advise ATC of specified information, e.g., "Report passing Hamilton VOR."

reporting accidents—(*See* AIRCRAFT ACCIDENT AND INCIDENT REPORTING.)

reporting point—A geographical location in relation to which the position of an aircraft is reported.

req—Request.

requirements for solo flight—(*See* SOLO FLIGHT REQUIREMENTS.)

rescue coordination center (RCC)—A primary search and rescue (SAR) facility suitably staffed by supervisory personnel and equipped for coodinating and controlling SAR operations in a region, subregion, or sector as defined by the national SAR plan.

restr—Restrict.

restricted area—Airspace of defined dimensions identified by an area on the surface of the earth within which the flight of aircraft, while not wholly prohibited, is subject to restrictions.

resume own navigation—Used by ATC to advise a pilot to resume his own navigational responsibility. It is issued after completion of a radar vector or when radar contact is lost while the aircraft is being radar vectored.

retesting after failure on the written test—Applicant may apply after first failure on presentation of a recommendation form from a certified instructor that he has been given additional instruction and is now ready for retesting. On second failure, applicant must wait 30 days.

rgd—Ragged.

rgt—Right.

rhumbline—The line drawn on a Lambert chart between points for navigational purposes. In practice it is the line on the map which the pilot attempts to follow.

rig—Adjustment of the airfoils of an airplane to produce desired flight characteristics.

right-of-way, converging aircraft at approximately the same altitude (except head on)—The aircraft to the other's right has the right-of-way. (*Also see* DISTRESS, HEAD-ON APPROACH, OVERTAKING.)

right-of way approaching airport—When two or more aircraft are approaching an airport for the purpose of landing, the aircraft at the lower altitude has the right-of-way, but it shall not take advantage of this rule to cut in front of another which is on final approach to land, or to overtake that aircraft.

right traffic indicator—At some airports, a flashing amber light is installed near the center of the segmented circle (but usually on top of the control tower or adjoining building) which indicates that a right traffic pattern is in effect at the time.

rime—(*See* GRAINY ICE.)

rmn—Remain.

RNAV—Random navigation.(*See* AREA NAVIGATION [RNAV] ROUTE SYSTEM.)

rnwy/rwy—Runway.

roger—Pilot's radio expression meaning I have received all of your last transmission. It should not be used to answer a question requiring a yes or no answer. (*See* AFFIRMATIVE, NEGATIVE.)

roll—Displacement around the longitudinal axis of an airplane.

rollout RVR—The RVR readout values obtained from RVR equipment located nearest the runway end.

rotating beacons—(*See* AIRPORT ROTATING BEACONS.)

route—A defined path, consisting of one or more courses, which an aircraft traverses in a horizontal plane over the surface of the earth.

route segment—As used in Air Traffic Control, a part of a route that can be defined by two navigational fixes, two NAVAIDs, or a fix and a NAVAID. (*See* FIX, ROUTE.)

rpd—Rapid.

rqr—Require.

RR—Low or medium frequency radio range station.

RRP—Runway reference point.

rte—Route.

rudder—A hinged, vertical, control surface used to induce or overcome yawing movements about the vertical axis.

rudder pedals (or bar)—Controls within the airplane by means of which the rudder is actuated.

ruf—Rough.

runway—A strip, either paved or improved, on which take-offs and landings are effected. Runways are numbered to correspond to their magnetic bearing. Runway 27, for example, has a bearing of 270 degrees. Surface wind direction issued by the tower is also magnetic. (Winds aloft are given in true degrees.)

runway edge light systems—Used to outline the edges of runways during periods of darkness and restricted visibility conditions; classified according to the intensity or brightness they are capable of producing. They are high intensity runway lights (HIRL), medium intensity runway lights (MIRL), and low intensity runway lights (LIRL). The HIRL and MIRL systems have variable intensity controls, whereas the LIRLs normally have one intensity setting.

runway end identifier lights (REIL)—Installed at many airfields to provide rapid and positive identification of the approach end of a particular runway. The system consists of a pair of synchronized flashing lights, one of which is located laterally on each side of the runway threshold facing the approach area. They are effective for: (1) identification of a runway surrounded by a preponderance of other lighting; (2) identification of a runway which lacks contrast with surrounding terrain; and (3) identification of a runway during reduced visibility.

runway marking—In the interest of safety, regularity, or efficiency of aircraft operations, the Federal Aviation Administration has recommended for the guidance of the public the following airport marking (runway numbers and letters are determined from the approach direction. The number is the whole number nearest one-tenth the magnetic azimuth of the centerline of the runway, measured clockwise from the magnetic north. For example, runway 34 is equivalent to 340 degrees. The letter or letters differentiate between parallel runways: For two parallel runways "L" "R"; for three parallel runways "L" "C" "R".):

1. Basic runway marking—used for operations under VFR: centerline marking and runway direction numbers (*See* Figure R-3).

2. Nonprecision instrument runway marking—Served by a nonvisual navigation aid and intended for landings under instrument weather conditions: basic runway markings plus threshold marking (*See* Figure R-4).

3. Precision instrument runway marking—served by nonvisual precision approach aids and on runways having special operational requirements, nonprecision instrument runway marking, touchdown zone marking, fixed distance marking, plus side stripes (*See* Figure R-5).

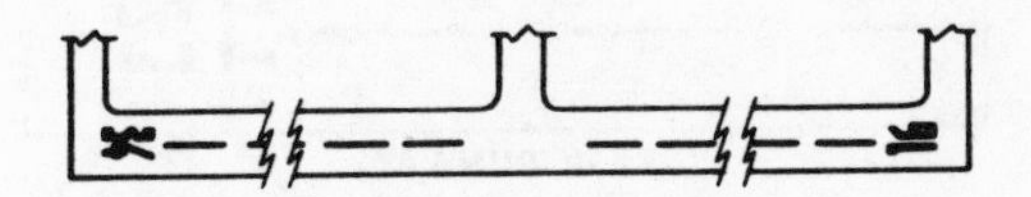

Figure R-3. Basic runway.

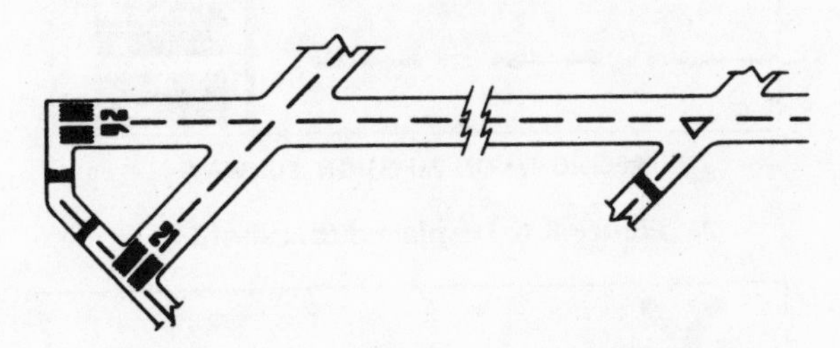

Figure R-4. Nonprecision instrument runway.

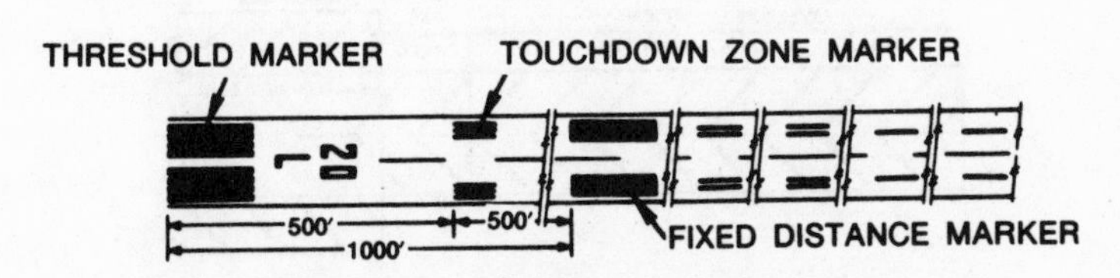

Figure R-5. Precision instrument runway.

runway marking *continued*

4. Threshold—A line perpendicular to the runway centerline designating the beginning of that portion of a runway usable for landing.

5. Displaced threshold—A threshold that is not at the beginning of the full strength runway pavement (*See* Figure R-6).

6. Closed or overrun/stopway areas—Any surface or area which appears usable but which, due to the nature of its structure, is unusable (*See* Figure R-7).

7. Fixed distance marker—To provide a fixed distance marker for landing of turbojet aircraft on other than a precision instrument runway. This marking is similar to the fixed distance marking on a precision instrument runway and located 1,000 feet from the threshold.

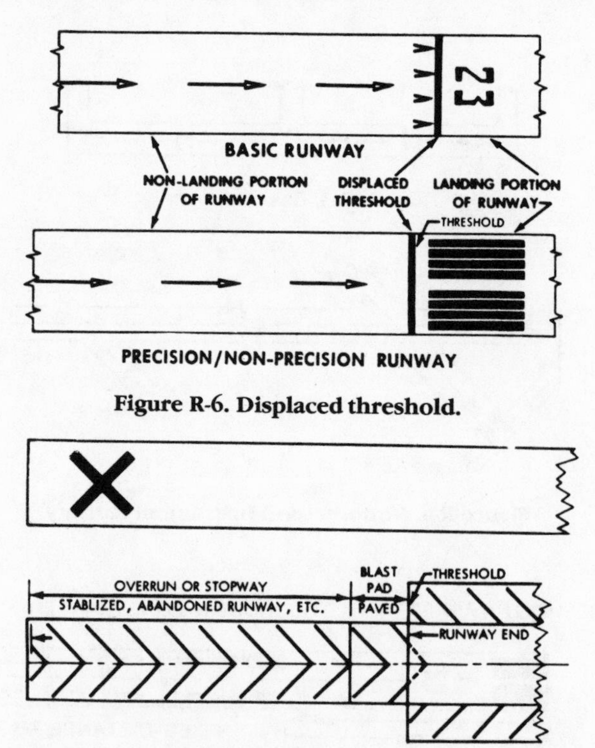

Figure R-6. Displaced threshold.

Figure R-7. Closed runway and overrun/stopway areas.

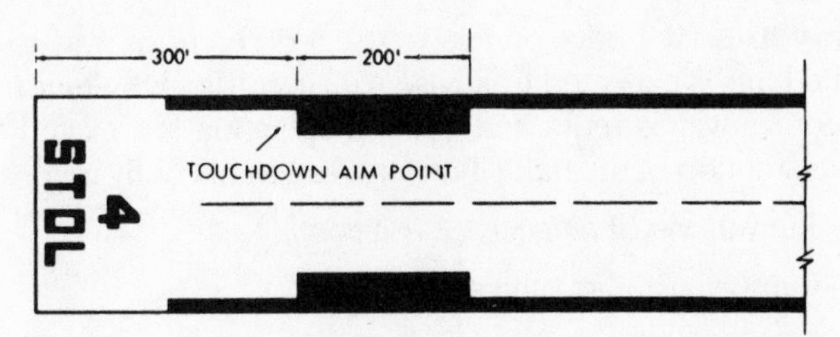

Figure R-8. STOL.

8. Short take-off and landing runway (STOL)—In addition to the normal runway number marking, the letters STOL are painted on the approach end of the runway and a touchdown aim point (*See* Figure R-8).

9. Taxiway marking—The taxiway centerline is marked with a continuous yellow line. The edges are marked with two continuous lines 6 inches apart. Taxiway HOLDING LINES consist of two continuous lines and two dashed lines perpendicular to the centerline. Pilots should stop short of the holding line for runup or when instructed by ATC to "HOLD SHORT OF (runway, ILS critical area, and so on)." Aircraft exiting a runway are not clear until the aircraft has passed the runway holding line.

10. Detailed airport marking information is published in Federal Aviation Administration advisory circular 150/5340-1, "Marking of Paved Areas On Airports." Advisory Circular 150/5390-1A, "Heliport Design Guide," contains details on heliport markings.

runway separation—Tower controllers establish the sequence of arriving and departing aircraft by requiring them to adjust flight or ground operation as necessary to achieve proper spacing. They may "HOLD" an aircraft short of the runway to achieve spacing between it and another arriving aircraft; the controller may instruct a pilot to "EXTEND DOWNWIND" in order to establish spacing from another arriving or departing aircraft. At times a clearance may include the word "IMMEDIATE." For example: "CLEARED FOR IMMEDIATE TAKEOFF." In such cases "IMMEDIATE" is used for purposes of air traffic separation. It is up to the pilot to refuse the clearance if, in his opinion, compliance would adversely affect his operation.

runway use—If a pilot prefers to use a different runway than that specified, he is expected to advise ATC accordingly. When use of a different runway is requested, pilot cooperation is solicited to preclude disruption of the traffic flow or creation of conflicting patterns.

RVR—Runway visual range. (*See* VISIBILITY.)

RVV—Runway visibility values.

rwy—Runway.

S

S in the phonetic alphabet is Sierra (see-air-ah).

S—A symbol for a superior air mass (warm dry air mass).

safe altitudes, minimum—(*See* MINIMUM SAFE ALTITUDE.)

safety advisory—A safety advisory issued by ATC to aircraft under their control if ATC is aware the aircraft is at an altitude which, in the controller's judgment, places the aircraft in unsafe proximity to terrain, obstructions, or other aircraft. The controller may discontinue the issuance of further advisories if the pilot advises he is taking action to correct the situation or has the other aircraft in sight.

1. Terrain/Obstruction Advisory—A safety advisory issued by ATC to aircraft under their control if ATC is aware the aircraft is at an altitude which, in the controller's judgment, places the aircraft in unsafe proximity to terrain/obstructions, e.g., "Low Altitude Alert, check your altitude immediately."

2. Aircraft Conflict Advisory—A safety advisory issued by ATC to aircraft under their control if ATC is aware of an aircraft that is not under their control at an altitude which, in the controller's judgment, places both aircraft in unsafe proximity to each other. With the alert, ATC will offer the pilot an alternate course of action when feasible, e.g., "Traffic Alert, advise you turn right heading zero niner zero or climb to eight thousand immediately."

The issuance of a safety advisory is contingent upon the capability of the controller to have an awareness of an unsafe condition. The course of action provided will be predicated on other traffic under

ATC control. Once the advisory is issued, it is solely the pilot's prerogative to determine what course of action, if any, he will take.

safety belts—Required for all occupants on all flights.

sailing—In seaplanes, the use of wind and current conditions to produce the desired track while taxiing on the water.

SAR—(*See* SEARCH AND RESCUE.)

say again—Pilot's radio expression used to request a repeat of the last transmission. Usually specifies transmission or portion thereof not understood or received, e.g., "Say again all after ABRAM VOR."

say altitude—Used by ATC to ascertain an aircraft's specific altitude/flight level. When the aircraft is climbing or descending, the pilot should state the indicated altitude rounded to the nearest 100 feet.

say heading—Used by ATC to request an aircraft's heading. The pilot should state the actual heading of the aircraft.

scatana plan—Security control of air traffic and air navigation aids.

sct—Scattered.

sctr—Sector.

scuba diving—If you fly to a sea resort or lake for a day's scuba diving, and then fly home, all within a few hours' time, this can be dangerous, particularly if you have been diving to depths for any length of time. Under the increased pressure of the water, excess nitrogen is absorbed into your system. If sufficient time has not elapsed prior to takeoff for your system to rid itself of this excess gas, you may experience the bends at altitudes under 8,000 feet where most light planes fly. Recommended waiting time is at least 4 hours after nondecompression diving; 24 hours after decompression diving.

SDF—simplified directional facility.

seaplane—An airplane equipped to rise from and alight on the water. Usually used to denote an airplane with detachable floats, as contrasted with a flying boat.

search and rescue (SAR)—A life-saving service provided through the combined efforts of the Federal Aviation Administration, Air Force, Coast Guard, State Aeronautic Commissions or other similar state agencies who are assisted by other organizations such as the Civil Air

search and rescue (SAR) *continued*

Patrol, Sheriffs' Air Patrol, State Police, and so on. It provides search, survival aid, and rescue of personnel of missing or crashed aircraft.

search and rescue facility—Responsible for maintaining and operating a search and rescue service to render aid to persons and property in distress.

second class medical certificate—(*See* MEDICAL CERTIFICATES.)

sectional chart—(*See* AERONAUTICAL CHARTS.)

see and avoid—A visual procedure wherein pilots of aircraft flying in visual meteorological conditions (VMC), regardless of type of flight plan, are charged with the responsibility to observe the presence of other aircraft and to maneuver their aircraft as required to avoid the other aircraft. Right-of-way rules are contained in FAR, Part 91.

segmented circle—A system designed to provide traffic pattern information at airports without operating control towers. (*See* Figure S-1.)

select code—That code displayed when the ground interrogator and the airborne transponder are operating on the same mode and code simultaneously.

separation—Spacing of aircraft to achieve their safe and orderly movement in flight and while landing and taking off.

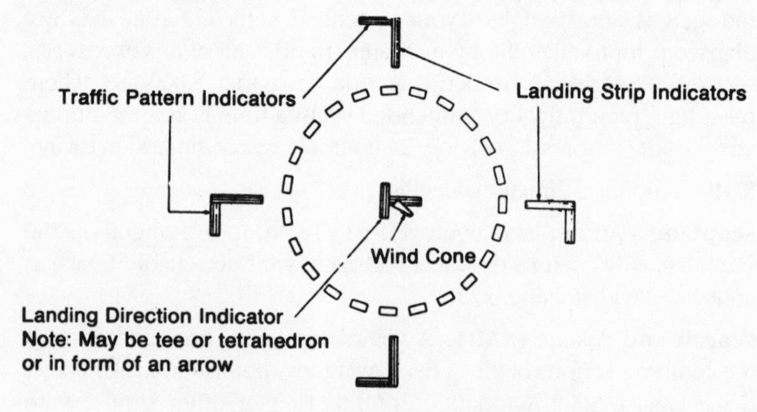

Figure S-1. Segmented circle.

separation minima—The minimum longitudinal, lateral, or vertical distances by which aircraft are spaced through the application of air traffic control procedures.

sequence report—The weather report transmitted hourly to all teletype stations, and available at all FSS communications stations.

servicing, aircraft at airports—(*See* AIRPORT SERVICING SYMBOLS.)

shft—Shift.

short takeoff and landing (STOL) aircraft—An aircraft which has the capability of operating from a STOL runway in accordance with applicable airworthiness and operating regulations.

short takeoff and landing (STOL) runway—A runway specifically designated and marked for STOL operations. (*See* RUNWAY MARKING.)

shwr—Shower.

SID—Standard instrument departure.

sideslip—(*See* SLIP.)

SIGMET (Significant meteorological information)—(Issued only if the hazardous weather was not accurately predicted in the area/terminal forecasts) include weather phenomena potentially hazardous to all aircraft, specifically: (1) tornadoes; (2) line of thunderstorms (squall lines); (3) embedded thunderstorms; (4) hail ¾″ or more; (5) severe and extreme turbulence; (6) severe icing; and (7) widespread duststorms/sandstorms, volcanic ash, lowering visibilities to less than 3 miles.

signals, light—(*See* LIGHT SIGNALS.)

simplified direction facility (SDF)—Provides a final approach course which is similar to that of the ILS localizer.

simulated instrument flight—Can only be undertaken in a plane with (1) dual controls; and (2) an appropriately rated pilot as safety pilot with adequate vision forward and to the side.

simultaneous landings on intersecting runways—The safety and operation of an aircraft are the responsibility of the pilot. If for *any* reason (difficulty in discerning location of intersection at night, wind factors, etc.) a pilot prefers to use the full length of the runway, a different runway, or desires to obtain the distance from the runway

simultaneous landings on intersecting runways ***continued***

threshold to the intersection, HE IS EXPECTED TO PROMPTLY INFORM ATC ACCORDINGLY.

simultaneous voice transmissions from a single FSS location—

1. At several Federal Aviation Administration facilities, simultaneous voice transmissions are made from a single FSS location. For example, the New York FSS transmits on the Kennedy, Hampton and Calverton VORTACs. If you are in the Calverton area, your callup should be "NEW YORK RADIO, CESSNA THREE ONE SIX ZERO FOXTROT, RECEIVING CALVERTON VOR, OVER."

2. If the chart indicates FSS frequencies above the VORTAC or in FSS communications boxes, transmit or receive on those frequencies nearest your location.

3. When unable to establish contact and you wish to call *any* ground station, use the phrase "ANY RADIO (tower) (station), GIVE CESSNA THREE ONE SIX ZERO FOXTROT A CALL ON (frequency) OR (VOR)." If an emergency exists or you need assistance, so state.

skc—Sky clear.

sked—Schedule.

skid—Sideward motion of an airplane in flight produced by centrifugal force; resulting from the rate of turn being too great for the angle of bank.

skill, aeronautical—(*See* PRIVATE PILOT, PREREQUISITES FOR CERTIFICATE.)

sky coverage symbols—(*See* Figure W-1.)

slgt—Slight.

slip—(or sideslip) The controlled flight of an airplane in a direction not in line with its longitudinal axis. Usually accomplished by holding aileron and rudder in opposed positions.

slipstream—The current of air driven astern by the propeller.

SM—Statute mile(s).

small aircraft—Usually aircraft of 12,500 pounds or less, maximum certified take-off weight.

smk—Smoke.

snw—Snow.

solo—A flight during which a pilot is the only occupant of the airplane.

solo cross-country flight, prerequisites for—Familiarity with flight planning, appropriate flight instruction, and endorsement of his student certificate to that effect.

solo flight, prerequisites for—Approved flight instruction, familiarity with general and visual flight rules of Part 91, Federal Aviation Regulations, and endorsement of his student certificate to that effect.

solo flight time—(*See* PRIVATE PILOT, PREREQUISITES FOR CERTIFICATE.)

space flight operations, flight limitations in proximity of—No person may operate any aircraft of U.S. registry or pilot any aircraft under the authority of an airman's certificate issued by the FAA within areas designated in a notice to airmen (NOTAM) for space flight operations except when authorized by ATC or operated under the control of the Dept. of Defense, Manager for Manned Space Flight Support Operations.

spar—The principal longitudinal structural member in an airfoil.

speak slower—Used in verbal communications as a request to reduce speech rate.

special emergency—

1. A condition of air piracy, or other hostile act by a person(s) aboard an aircraft, which threatens the safety of the aircraft or its passengers.

2. The pilot of an aircraft reporting a special emergency condition should: (a) (if circumstances permit) apply distress or urgency radiotelephone, including the details of the condition; (b) (if circumstances do not permit the use of prescribed distress or urgency procedures) send the message on the air-ground frequency in use at the time, consisting of as many as possible of the following elements spoken distinctly and in the following order: name of the station addressed (time and circumstances permitting), the identification of the aircraft and present position, and the nature of the special emergency condition and pilot intentions (circumstances permitting). (If unable to provide the latter, use code words and/or transponder setting for indicated meanings as follows: spoken words: TRANSPONDER SEVEN FIVE ZERO ZERO; meaning: am being hijacked/

special emergency *continued*

forced to a new destination; transponder setting: mode 3/A, Code 7500.)

3. Air traffic controllers will acknowledge and confirm the receipt of transponder Code 7500 by asking the pilot if the code is intentionally being used. If the pilot replies in the affirmative or does not reply, the controller will not ask further questions but will follow flight, respond to pilot's requests, and notify appropriate authorities.

special use airspace—Consists of that airspace wherein activities must be confined because of their nature, or wherein limitations are imposed upon aircraft operations that are not a part of those activities, or both. These areas are depicted on aeronautical charts.

special VFR clearances—(*See* VFR CLEARANCES, SPECIAL.)

speed adjustment—An ATC procedure used to request pilots to adjust aircraft speed to a specific value for the purpose of providing desired spacing. Pilots are expected to maintain a speed of plus or minus 10 knots or 0.02 mach number of the specified speed.

speed, aircraft in airport traffic areas—(*See* AIRPORT, OPERATION IN VICINITY OF.)

speeds, radio phraseology—The separate digits of the speed followed by the word knots. The controller may omit the word knots when using speed adjustment procedures, "Reduce/Increase Speed to One Five Zero." Examples: 250—TWO FIVE ZERO KNOTS; 185—ONE EIGHT FIVE KNOTS; 95—NINER FIVE KNOTS.

spin—A prolonged stall in which an airplane rotates about its center of gravity while it descends, usually with its nose well down.

spiral—A prolonged gliding or climbing turn during which at least 360 degrees change of direction is effected.

spot landing—(*See* ACCURACY LANDINGS.)

sqln—Squall line.

SQUAWK (Mode, Code, Function)—Activate specific modes/codes/functions on the aircraft transponder, e.g., "Squawk three/alpha, two one zero five, low." (*See* TRANSPONDER.)

SR—Sunrise.

SS—Sunset.

stability—The tendency of an airplane in flight to remain in straight, level, upright position, or to return to this attitude if displaced, without attention of the pilot.

stabilizer—The fixed airfoil of an airplane used to increase stability; usually, the aft fixed horizontal surface to which the elevators are hinged.

stage I/II/III service—(*See* TERMINAL RADAR PROGRAM.)

stall—The loss of lift when the air speed decreases to the minimum which will support an airfoil at the existing loading.

stand by—Radio expression means the controller or pilot must pause for a few seconds, usually to attend to other duties of a higher priority. Also means to wait as in "stand by for clearance." If a delay is lengthy, the caller should reestablish contact.

standard atmosphere—The air is a dry perfect gas; temperature at sea level is 59 degrees Fahrenheit, pressure at sea level is 29.92 in. Hg.

standard instrument departure (SID)—A preplanned coded air traffic control IFR departure routing, preprinted for pilot use in graphic and textual or textual form only.

standard rate turn—A turn of 3 degrees per second.

standard terminal arrival route (STAR)—A preplanned coded air traffic control IFR arrival routing, established for application to arriving aircraft destined for certain airports, and preprinted for pilot use in graphic and textual or textual form only. Its purpose is to simplify clearance delivery procedures.

STAR—Standard terminal arrival route.

statute mile—5,280 feet. (*Also see* NAUTICAL MILE.)

stbl—Stable.

step—A "break" in the bottom of a seaplane's float to improve planing characteristics on the water.

stg—Strong.

STOL—Short takeoff and landing.

stop altitude squawk—Expression used by ATC to inform an aircraft to turn-off the automatic altitude reporting feature of its transponder. It is issued when the verbally reported altitude varies 300 feet or more from the automatic altitude report.

stop squawk (Mode or Code)—Used by ATC to inform an aircraft to turn-off the automatic altitude reporting feature of its transponder. It is issued when the verbally reported altitude varies 300 feet or more from the automatic altitude report.

stopway—Area beyond take-off runway used in decelerating airplane during an aborted takeoff.

straight-in approach (VFR)—Entry into traffic pattern by interception of the extended runway centerline without executing any other portion of the normal traffic pattern.

stratus—(*See* CLOUDS.)

strut—A compression or tension member in a truss structure. In airplanes, usually applied to an external major structural member.

stuck mike—Be alert to the sounds *or lack of sounds* in your receiver. Check your volume, recheck your frequency and *make sure that your microphone is not stuck* in the transmit position. Frequency blockage can, and has, occurred for extended periods of time due to unintentional transmitter operation. This type of interference is commonly referred to as a "stuck mike," and controllers may refer to it in this manner when attempting to assign an alternate frequency. If the assigned frequency is completely blocked by this type of interference, use the procedures described in the *Airman's Information Manual* for en route IFR radio frequency outage to establish or reestablish communications with ATC.

student pilot certificate, duration—Certificate expires at the end of the 24th month from the date it is issued.

student pilot certificate, prerequisites for—16 years of age, ability to speak, read, and understand English language, and a third-class medical certificate.

student pilot limitations—A student pilot may not act in command of an aircraft except under specific rules as listed in Federal Aviation Regulations Part 61.

student pilot medical certificate—A third class medical certificate is required of student pilots.

student pilot minimum age—16 years.

student pilot's radio identification—The Federal Aviation Administration desires to help the student pilot in acquiring sufficient practical experience in the environment in which he will be required to operate. To receive additional assistance while operating in areas of concentrated air traffic, a student pilot need only identify himself as a student pilot during his initial call to a Federal Aviation Administration radio facility. (Example: "Dayton Tower, this is Fleetwing November 1234, Student Pilot, over.") This special identification will alert Federal Aviation Administration air traffic control personnel and enable them to provide the student pilot with such extra assistance and consideration as he may need. This procedure is not mandatory.

surveillance approach—An instrument approach conducted in accordance with directions issued by a controller referring to the surveillance radar display.

surveillance radar—

1. Surveillance radars are divided into two general categories: airport surveillance radar (ASR) and air route surveillance radar (ARSR). ASR is designed to provide relatively short-range coverage in the general vicinity of an airport and to serve as an expeditious means of handling terminal area traffic through observation of precise aircraft locations on a radarscope. The ASR can also be used as an instrument approach aid. ARSR is a long-range radar system designed primarily to provide a display of aircraft locations over large areas.

2. Surveillance radars scan through 360 degree of azimuth and present target information on a radar display located in a tower or center. This information is used independently or in conjunction with other navigational aids in the control of air traffic.

survival radio equipment—A self-buoyant, water resistant, portable emergency radio signaling device which operates from its own power source on 121.5 and/or 243 MHz, preferably on both emergency frequencies, transmitting a distinctive downward swept audio tone for homing purposes, which may or may not have voice capability, and which is capable of operation by unskilled persons. This type of equipment is agreed upon internationally for extended overwater operations and is presently required for air carriers engaged in extended overwater operations.

svc—Service.

svr—Severe.

symbols—

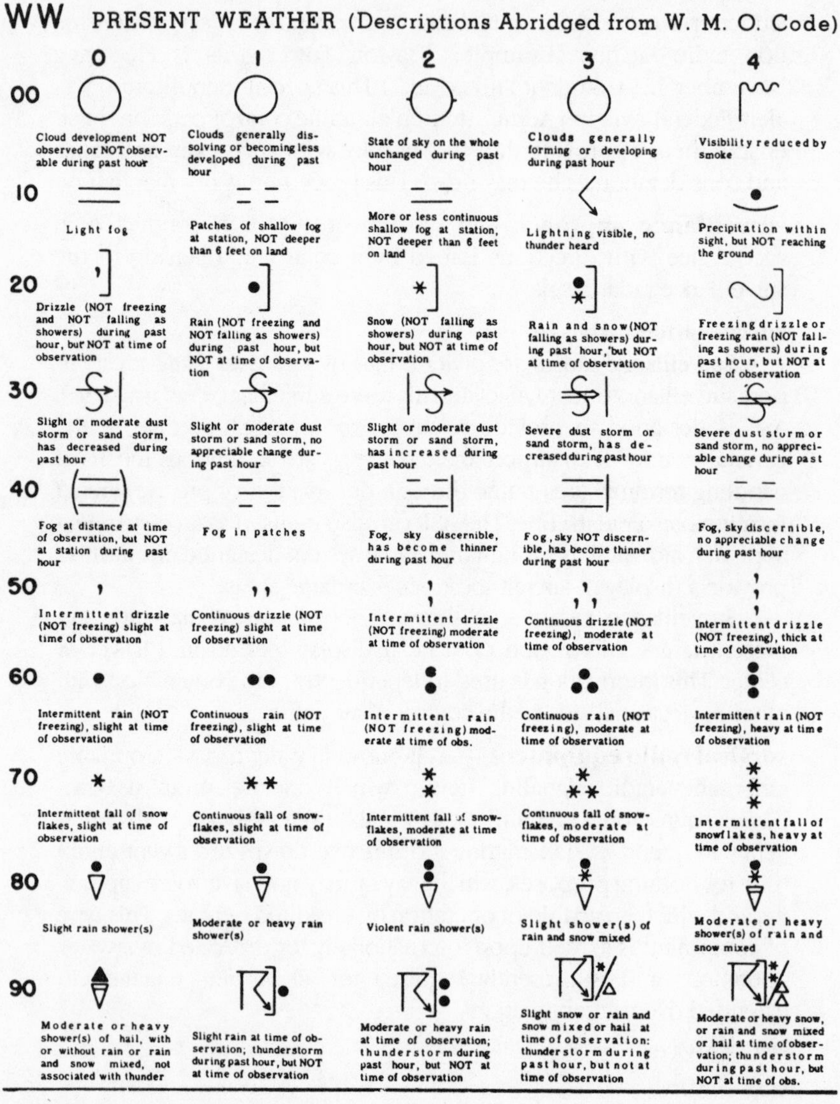

WW PRESENT WEATHER (Descriptions Abridged from W. M. O. Code)

	0	1	2	3	4
00	Cloud development NOT observed or NOT observable during past hour	Clouds generally dissolving or becoming less developed during past hour	State of sky on the whole unchanged during past hour	Clouds generally forming or developing during past hour	Visibility reduced by smoke
10	Light fog	Patches of shallow fog at station, NOT deeper than 6 feet on land	More or less continuous shallow fog at station, NOT deeper than 6 feet on land	Lightning visible, no thunder heard	Precipitation within sight, but NOT reaching the ground
20	Drizzle (NOT freezing and NOT falling as showers) during past hour, but NOT at time of observation	Rain (NOT freezing and NOT falling as showers) during past hour, but NOT at time of observation	Snow (NOT falling as showers) during past hour, but NOT at time of observation	Rain and snow (NOT falling as showers) during past hour, but NOT at time of observation	Freezing drizzle or freezing rain (NOT falling as showers) during past hour, but NOT at time of observation
30	Slight or moderate dust storm or sand storm, has decreased during past hour	Slight or moderate dust storm or sand storm, no appreciable change during past hour	Slight or moderate dust storm or sand storm, has increased during past hour	Severe dust storm or sand storm, has decreased during past hour	Severe dust storm or sand storm, no appreciable change during past hour
40	Fog at distance at time of observation, but NOT at station during past hour	Fog in patches	Fog, sky discernible, has become thinner during past hour	Fog, sky NOT discernible, has become thinner during past hour	Fog, sky discernible, no appreciable change during past hour
50	Intermittent drizzle (NOT freezing) slight at time of observation	Continuous drizzle (NOT freezing) slight at time of observation	Intermittent drizzle (NOT freezing) moderate at time of observation	Continuous drizzle (NOT freezing), moderate at time of observation	Intermittent drizzle (NOT freezing), thick at time of observation
60	Intermittent rain (NOT freezing), slight at time of observation	Continuous rain (NOT freezing), slight at time of observation	Intermittent rain (NOT freezing) moderate at time of obs.	Continuous rain (NOT freezing), moderate at time of observation	Intermittent rain (NOT freezing), heavy at time of observation
70	Intermittent fall of snow flakes, slight at time of observation	Continuous fall of snowflakes, slight at time of observation	Intermittent fall of snowflakes, moderate at time of observation	Continuous fall of snowflakes, moderate at time of observation	Intermittent fall of snowflakes, heavy at time of observation
80	Slight rain shower(s)	Moderate or heavy rain shower(s)	Violent rain shower(s)	Slight shower(s) of rain and snow mixed	Moderate or heavy shower(s) of rain and snow mixed
90	Moderate or heavy shower(s) of hail, with or without rain or rain and snow mixed, not associated with thunder	Slight rain at time of observation; thunderstorm during past hour, but NOT at time of observation	Moderate or heavy rain at time of observation; thunderstorm during past hour, but NOT at time of observation	Slight snow or rain and snow mixed or hail at time of observation: thunderstorm during past hour, but not at time of observation	Moderate or heavy snow, or rain and snow mixed or hail at time of observation; thunderstorm during past hour, but NOT at time of obs.

Figure S-2. Weather symbols.

5	6	7	8	9
Haze	Widespread dust in suspension in the air, NOT raised by wind, at time of observation	Dust or sand raised by wind, at time of observation	Well developed dust devil(s) within past hour	Dust storm or sand storm within sight of or at station during past hour
Precipitation within sight, reaching the ground, but distant from station	Precipitation within sight, reaching the ground, near to but NOT at station	Thunder heard, but no precipitation at the station	Squall(s) within sight during past hour	Funnel cloud(s) within sight during past hour
Showers of rain during past hour, but NOT at time of observation	Showers of snow, or of rain and snow, during past hour, but NOT at time of observation	Showers of hail, or of hail and rain, during past hour, but NOT at time of observation	Fog during past hour, but NOT at time of observation	Thunderstorm (with or without precipitation) during past hour, but NOT at time of obs.
Severe dust storm or sand storm, has increased during past hour	Slight or moderate drifting snow, generally low	Heavy drifting snow, generally low	Slight or moderate drifting snow, generally high	Heavy drifting snow, generally high
Fog, sky NOT discernible, no appreciable change during past hour	Fog, sky discernible, has begun or become thicker during past hour	Fog, sky NOT discernible, has begun or become thicker during past hour	Fog, depositing rime, sky discernible	Fog, depositing rime, sky NOT discernible
Continuous drizzle (NOT freezing), thick at time of observation	Slight freezing drizzle	Moderate or thick freezing drizzle	Drizzle and rain, slight	Drizzle and rain, moderate or heavy
Continuous rain (NOT freezing), heavy at time of observation	Slight freezing rain	Moderate or heavy freezing rain	Rain or drizzle and snow, slight	Rain or drizzle and snow, moderate or heavy
Continuous fall of snowflakes, heavy at time of observation	Ice needles (with or without fog)	Granular snow (with or without fog)	Isolated starlike snow crystals (with or without fog)	Ice pellets (sleet, U. S. definition)
Slight snow shower(s)	Moderate or heavy snow shower(s)	Slight shower(s) of soft or small hail with or without rain, or rain and snow mixed	Moderate or heavy shower(s) of soft or small hail with or without rain, or rain and snow mixed	Slight shower(s) of hail, with or without rain or rain and snow mixed, not associated with thunder
Slight or moderate thunderstorm without hail, but with rain and/or snow at time of obs.	Slight or moderate thunderstorm, with hail at time of observation	Heavy thunderstorm, without hail, but with rain and/or snow at time of observation	Thunderstorm combined with dust storm or sand storm at time of obs.	Heavy thunderstorm with hail at time of observation

sys—System.

T

T in the phonetic alphabet is Tango (tan-go).

T—Symbol for tropical air mass.

T—True (after a bearing).

tab—A small auxiliary airfoil, usually attached to a movable control surface to aid in its movement, or to effect a slight displacement of it for the purpose of trimming the airplane for varying conditions of power, load, or airspeed.

TACAN—Tactical air navigation.

tachometer—An instrument which shows, in revolutions per minute (RPM), the rate at which the engine crankshaft is revolving.

tactical air navigation (TACAN)—UHF navigational facility, omnidirectional course and distance information. For reasons peculiar to military or naval operations (unusual siting conditions, the pitching and rolling of a naval vessel, etc.) the civil VOR-DME system of air navigation was considered unsuitable for military or naval use. A new navigational system, TACAN, was therefore developed by the military and naval forces to more readily lend itself to military and naval requirements. As a result, the Federal Aviation Administration has been in the process of integrating TACAN facilities with the civil VOR-DME program. Although the theoretical, or technical principles, of operation of TACAN equipment are quite different from those of VOR-DME facilities, the end result, as far as the navigating pilot is concerned, is the same. These integrated facilities are called VORTACs.

tail group—The airfoil members of the assembly located at the rear of an airplane.

tailheavy—A condition of trim in an airplane in which the tail tends to sink.

tailskid—A skid, or runner, which supports the aft end of the airframe while on the ground.

tail slide—Rearward motion of an airplane in the air; commonly occurs only in a whip stall.

tail wheel—A wheel which serves the same purpose as the tailskid.

take-off clearance—Tower controllers establish the sequence of arriving and departing aircraft by requiring them to adjust flight or ground operation as necessary to achieve proper spacing. They may "HOLD" an aircraft short of the runway to achieve spacing between it and another arriving aircraft. At times a clearance may include the word "IMMEDIATE." For example: "CLEARED FOR IMMEDIATE TAKE-OFF." In such cases "IMMEDIATE" is used for purposes of air traffic separation. It is up to the pilot to refuse the clearance if, in his opinion, compliance would adversely affect his operation.

takeoffs, intersection—(*See* INTERSECTION TAKEOFFS.)

taxi—To operate an airplane under its own power on the ground, except that movement incident to actual takeoff and landing.

taxi clearance—At controlled airports, must be obtained before moving the aircraft by calling ground control on the frequency listed for that airport in the *Airport/Facility Directory.*

taxi into position and hold—Used by ATC to inform a pilot to taxi onto the departure runway in takeoff position and hold. It is not authorization for takeoff. It is used when take-off clearance cannot immediately be issued because of traffic or other reasons.

taxiway marking—The taxiway centerline is marked with a continuous yellow line. The taxiway edge may be marked with two continuous yellow lines 6 inches apart. Taxiway holding lines consist of two continuous and two dashed lines spaced 6 inches between lines, perpendicular to the centerline.

TCA—Terminal control area.

TCH—Threshold crossing height.

temporary flight restrictions—

1. Temporary flight restrictions may be put into effect in the vicinity of any incident or event which by its nature may generate such a high degree of public interest that the likelihood of a hazardous congestion of air traffic exists. Federal Aviation Regulation Part 91 prohibits the operation of nonessential aircraft in airspace that has been designated in a NOTAM as an area within which temporary flight restrictions apply. The revised rule will continue to be implemented in the case of disasters of substantial magnitude. It will also be implemented as necessary in the case of demonstrations, riots, and other civil disturbances, as well as major sporting events, parades, pageants, and similar functions which are likely to attract large crowds and encourage viewing from the air.

2. NOTAM's implementing temporary flight restrictions will contain a description of the area in which the restrictions apply. Normally the area will include the airspace below 2,000 feet above the surface within 5 miles of the site of the incident. However, the exact dimensions will be included in the NOTAM.

temporary pilot certificate—Issued to an applicant for not more than 120 days pending the issue of the actual certificate or rating for which he applied.

terminal area charts—(*See* AERONAUTICAL CHARTS.)

terminal chart figures—Such as $\frac{70}{30}$ indicate 7,000 and 3,000 tops and bottoms of controlled airspace.

terminal control area (TCA)—Consists of controlled airspace extending upward from the surface or higher to specified altitudes, within which all aircraft are subject to operating rules and pilot and equipment requirements specified in Part 91 of the Federal Aviation Regulations. TCAs are described in Part 71 of the Federal Aviation Regulations. Each such location is designated as a Group I or Group II terminal control area, and includes at least one primary airport around which the TCA is located. (*See* Federal Aviation Regulations, Part 71.)

1. Group I terminal control areas represent some of the busiest locations in terms of aircraft operations and passengers carried, and it is necessary for safety reasons to have stricter requirements for operation within Group I TCAs. Student pilots are forbidden.

2. Group II terminal control areas represent less busy locations,

and though safety dictates some pilot and equipment requirements, they are not as stringent as those for Group I locations. Student pilots are permitted.

3. As terminal control areas come into being, they will be shown on sectional and other charts such as VFR TERMINAL AREA CHARTS.

terminal control area operation—The operating rules and pilot and equipment requirements for operating within a terminal control area (TCA) are found in Federal Aviation Regulations, Part 91. Pilots should not request a clearance to operate within a TCA unless these requirements are met. (*See* TERMINAL CONTROL AREA.)

1. Operating rules and pilot/equipment requirements—Regardless of weather conditions, or whether the pilot is on an IFR flight plan or VFR, an ATC authorization is required prior to operating within a TCA. Additional requirements include: (a) a two-way radio capable of communicating with ATC on appropriate frequencies; (b) a VOR or TACAN receiver; (c) an appropriate radar beacon transponder; and (d) private pilot certificate or better in order to land or take off from an airport within a Group I TCA.

2. Flight procedures—VFR Flights: (a) arriving VFR flights should contact ATC on the appropriate frequency and in relation to geographical fixes shown on local charts. Although a pilot may be operating beneath the floor of the TCA on initial contact, communications with ATC should be established in relation to the points indicated for spacing and sequencing purposes; (b) departing VFR aircraft should advise clearance delivery of the intended altitude and route of flight to depart the TCA; (c) aircraft not landing/departing the primary airport may obtain ATC clearance to transit the TCA when traffic conditions permit and provided the requirements of Federal Aviation Regulation Part 91 are met. Such VFR transiting aircraft are encouraged, to the extent possible, to transit through VFR corridors or above or below the TCA.

Pilots operating in VFR corridors are urged to use frequency 122.750 MHz for the exchange of aircraft position information. VFR non-TCA aircraft are cautioned against operating too closely to TCA boundaries, especially where the floor of the TCA is 3,000 feet or less or where normal VFR cruise altitudes are at or near the floor of higher levels. Observance of this precaution will reduce the potential for encountering a TCA aircraft operating at TCA floor altitudes.

terminal forecasts—Contain information for specific airports on expected ceiling, cloud heights, cloud amounts, visibility, weather and obstructions to vision and surface wind. They are issued three times/day and are valid for 24 hours.

terminal radar programs for VFR aircraft—A national program instituted to extend the terminal radar services, provided IFR aircraft, to VFR aircraft. Pilot participation in the program is urged but is not mandatory. The program is divided into two parts and referred to as Stage II and Stage III. The Stage service provided at a particular location is contained in *Airport/Facility Directory.*

1. Stage I originally comprised two basic radar services (traffic advisories and limited vectoring to VFR aircraft). These services are provided by all commissioned terminal radar facilities, but the term "Stage I" has been deleted from use.

2. Stage II service (radar advisory and sequencing service for VFR aircraft): (a) has been implemented at certain terminal locations. Pilot participation is urged but is not mandatory; (b) its purpose is to adjust the flow of arrival VFR and IFR aircraft into the traffic pattern in a safe and orderly manner. Pilots of arriving VFR aircraft should initiate radio contact (frequencies listed in the *Airport/Facility Directory*) with approach control when approximately 25 miles from the airport at which the Stage II services are provided, and state: "REQUEST RADAR SERVICE." Approach control then provides the pilot with wind and runway, except when the pilot states "HAVE NUMBERS" or the information is contained in the ATIS broadcast and the pilot indicates that he has received the ATIS information, routings, etc., as necessary for proper sequencing with other traffic en route to the airport. Traffic information will be provided on a workload-permitting basis; (c) after radar contact is established, the pilot is directed to fly specific headings either to join an appropriate leg of the traffic pattern or to position the flight behind the preceding aircraft in the approach sequence. When a flight is positioned behind the preceding aircraft and the pilot reports having that aircraft in sight, he will be directed to follow it. Upon being told to contact the tower, radar service is automatically terminated. A landing sequence number will thereafter be assigned by the tower, unless it was previously issued by approach control. Standard radar separation will be provided between IFR aircraft until

such time as the aircraft is sequenced and the pilot sees the traffic he is to follow. Standard radar separation between VFR or between VFR and IFR aircraft will not be provided; (d) pilots of departing VFR aircraft desiring traffic information service should state: "REQUEST RADAR SERVICE" on initial contact with ground control and advise proposed direction of flight. Following takeoff, the tower will advise when to contact departure control and the frequency to be used; (e) pilots of aircraft transiting the area may receive traffic information on a controller workload-permitting basis. Those desiring such service should state: "REQUEST RADAR TRAFFIC INFORMATION," give their position, altitude, radar beacon code (if transponder equipped), destination, and/or route of flight.

3. Stage III (radar sequencing and separation service for VFR aircraft:

a. It has been implemented at certain terminal locations and is provided on a voluntary pilot participating basis.

b. Its purpose is to provide, to the extent possible, separation between all participating VFR aircraft and all IFR aircraft operating within the airspace defined as the TRSA (terminal radar service area).

c. The TRSA is primarily a radar environment and control is predicated thereon; however, this does not preclude application of non-radar separation when required or deemed appropriate. During weather conditions equal to or better than basic VFR, 500 feet vertical separation may be used between VFR aircraft and/or between VFR and IFR aircraft, or visual separation may be used between arriving VFR aircraft and/or between VFR and IFR aircraft as follows:

(i) When a VFR flight is positioned behind the preceding aircraft and the pilot reports having that aircraft in sight, he will be directed to follow it;

(ii) When IFR flights, operating in VFR conditions, are being sequenced with other traffic, and the pilot reports the aircraft he is to follow in sight, the pilot may be directed to follow it and may be cleared for a "visual approach;"

(iii) Upon being told to contact the tower, radar service is automatically terminated. The tower will assign a landing sequence number, unless it was previously issued by approach control.

Visual separation may be used between departing and arriving VFR

terminal radar programs for VFR aircraft *continued*

aircraft and/or between VFR and IFR aircraft when the pilot sees the other aircraft and reports he will maintain visual separation from it.

(iv) Within the TRSA, traffic information on observed but unidentified targets will, to the extent possible, be provided all IFR and participating VFR aircraft, and such aircraft, at the request of the pilot, will be vectored to avoid the observed traffic, insofar as possible, provided the aircraft to be vectored is within the airspace under the jurisdiction of the controller.

4. Pilots responsibility: These programs are not to be interpreted as relieving pilots of their responsibilities to see and avoid other traffic operating in basic VFR weather conditions, to maintain appropriate terrain and obstruction clearance, or to remain in weather conditions equal to or better than the minima required by Federal Aviation Regulation Part 91. Whenever compliance with an assigned route or heading is likely to compromise said pilot responsibility respecting terrain and obstruction clearance and weather minima, approach control should be so advised and a revised clearance or instruction obtained.

terminal radar service area (TRSA)—Airspace surrounding designated airports wherein ATC provides radar vectoring, sequencing, and separation on a fulltime basis for all IFR and participating VFR aircraft. Service provided in a TRSA is called Stage III Service. AIM contains an explanation of TRSA. TRSAs are depicted on VFR aeronautical charts. Pilot participation is urged but is not mandatory. (*See* TERMINAL RADAR PROGRAM.)

terminal velocity—The hypothetical maximum speed which could be obtained in a prolonged vertical dive.

tetrahedron—A device normally located on uncontrolled airports and used as a landing direction indicator. The small end of a tetrahedron points in the direction of landing. At controlled airports, the tetrahedron, if installed, should be disregarded because tower instructions supersede the indicator.

tetrahedron, lighted—During hours of darkness, flashing lights outlining the tetrahedron or wind tee means that ground visibility is less than 3 miles and/or that the ceiling is less than 1,000 feet. Not all traffic indicators are equipped with flashing lights. This should not be the only means used to determine weather conditions.

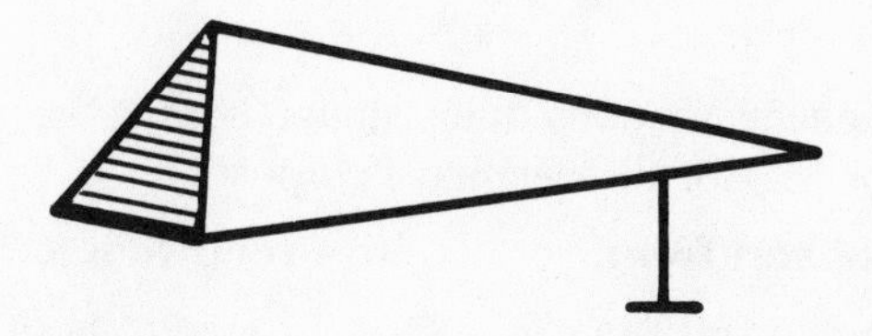

Figure T-1. Tetrahedron. (NOTE: Direction of landing → .)

tfc—Traffic.

third class medical certificate—(*See* MEDICAL CERTIFICATES.)

thld—Threshold.

thn—Thin.

threshold—The beginning of that portion of the runway usable for landing.

threshold crossing height (TCH)—The height of the straight line extension of the visual or electronic glide slope above the runway threshold.

thrust—The forward force on an airplane in the air, provided by the engine acting through a propeller in conventional airplanes.

thunderstorms—The visible thunderstorm cloud is only a portion of a turbulent system whose updrafts and downdrafts often extend far beyond the visible storm cloud. Severe turbulence can be expected up to 20 miles from severe thunderstorms; this distance decreases to about 10 miles in less severe storms. These turbulent areas may appear as a well defined echo on weather radar. Turbulence beneath a thunderstorm should not be minimized. This is especially true when the relative humidity is low in any layer between the surface and 15,000 feet. Then the lower altitudes may be characterized by strong out-flowing winds and severe turbulence. The probability of lightning strikes occurring to aircraft is greatest when operating at altitudes where temperatures are between -5 degrees Centigrade and +5 degrees Centigrade. Lightning can strike aircraft flying in the clear in the vicinity of a thunderstorm. A "soft" altitude in one thunderstorm may be a most severe altitude in another. While it is impossible to state rules for safe flight through a thunderstorm, there is one sure caution: STAY OUT OF THEM AND GIVE THEM A WIDE BERTH.

til—Until.

time—The Federal Aviation Administration utilizes Greenwich mean time (GMT or "Z") for all operational purposes.

To convert from:	to Greenwich mean time:
Eastern Standard Time	add 5 hours
Eastern Daylight Time	add 4 hours
Central Standard Time	add 6 hours
Central Daylight Time	add 5 hours
Mountain Standard Time	add 7 hours
Mountain Daylight Time	add 6 hours
Pacific Standard Time	add 8 hours
Pacific Daylight Time	add 7 hours

time—Phraseology in radio communications: the 24-hour clock system is used in radiotelephone transmissions. The hour is indicated by the first 2 figures and the minutes by the last 2 figures. Examples: 0000—ZERO ZERO ZERO ZERO; 0920—ZERO NINER TWO ZERO. Time may be stated in minutes only (two figures) in radiotelephone communications when no misunderstanding is likely to occur. Current time in use at a station is stated in the nearest ¼-minute in order that pilots may use this information for time checks. Fractions of a ¼-minute less than 8 seconds are stated as the preceding ¼-minute; fractions of a ¼-minute of 8 seconds or more are stated as the succeeding ¼-minute. Examples:

0929:05—TIME, ZERO NINER TWO NINER
0929:10—TIME, ZERO NINER TWO NINER AND ONE-QUARTER
0929:28—TIME, ZERO NINER TWO NINER AND ONE-HALF.

tkf—Takeoff.

tmprly—Temporarily.

tmpry—Temporary.

topographical symbols on charts—(*See* Figure T-2.)

torching—The burning of fuel at the end of an exhaust pipe or stack of a reciprocating aircraft engine, the result of an excessive richness in the fuel air mixture.

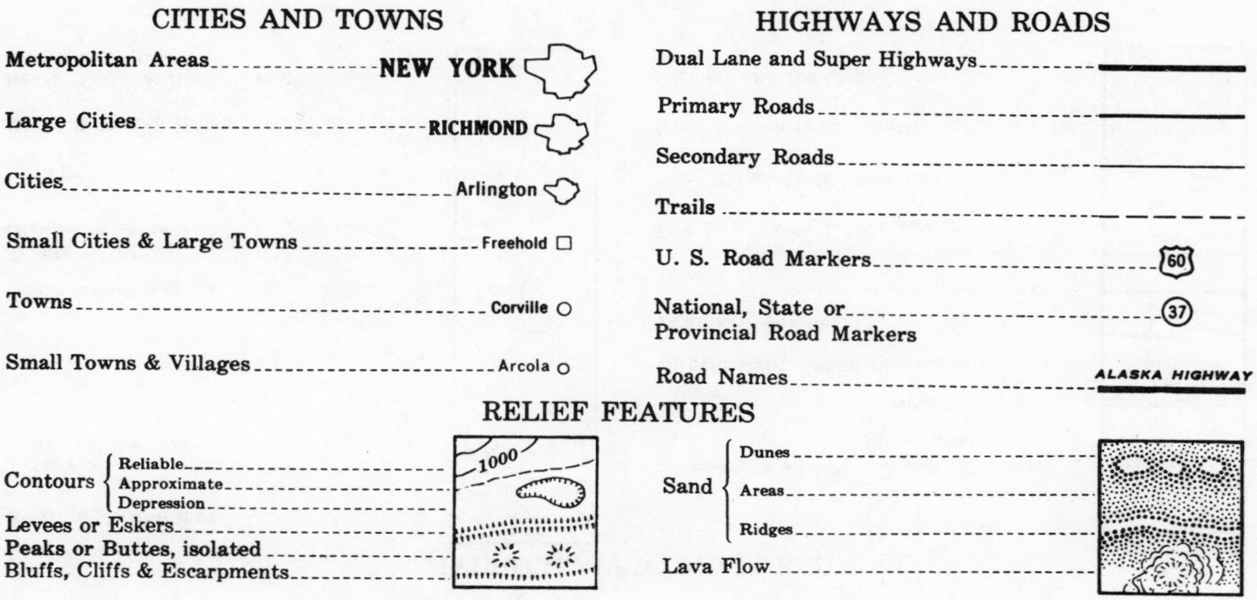

CITIES AND TOWNS

Metropolitan Areas ---------- NEW YORK

Large Cities ---------- RICHMOND

Cities ---------- Arlington

Small Cities & Large Towns ---------- Freehold

Towns ---------- Corville

Small Towns & Villages ---------- Arcola

HIGHWAYS AND ROADS

Dual Lane and Super Highways

Primary Roads

Secondary Roads

Trails

U. S. Road Markers ---------- 60

National, State or Provincial Road Markers ---------- 37

Road Names ---------- ALASKA HIGHWAY

RELIEF FEATURES

Contours { Reliable / Approximate / Depression

1000

Levees or Eskers

Peaks or Buttes, isolated

Bluffs, Cliffs & Escarpments

Sand { Dunes / Areas / Ridges

Lava Flow

Figure T-2. Topographical symbols.

HYDROGRAPHIC FEATURES

Swamps & Marshes

Tidal Flats
(Exposed at low tide)

Danger Line

Rocks Awash

Shoals
(Exposed at low tide)

Springs

Wells & Water Holes

Reefs, Coral & Rocky Ledges
(Awash at low tide)

Streams & Rivers
- Perennial
- Intermittent
- Probable or Unsurveyed
- Braided

Intermittent Lakes (blue stipple)

Drainage Ditches

Canals
- In use
- Abandoned

Dry Lake Beds (brown stipple)

Sand Deposits in river bed

Dry Washes (brown stipple)

Glaciers and Ice Caps

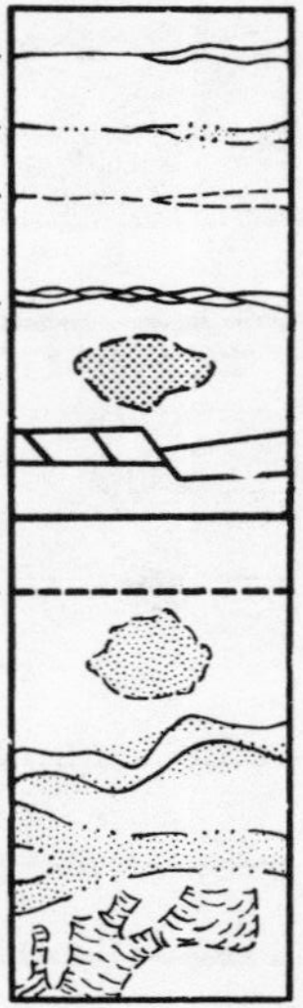

CULTURAL AND MISCELLANEOUS

Landmarks (with appropriate note) (Numerals indicate elevation above sea level of top)	■ Factory ■ Stack 875′
Oil Tanks	
Oil Fields	
Dams	
Rapids and Falls	
Elevations (In feet) — Highest on chart	•1115
Elevations (In feet) — Highest in a general area	•1085
Elevations (In feet) — Spot	•950
Mines and Quarries	
Mountain Passes	
Lookout Stations (Elevation is base of tower)	75 (Site) 1025 (Elev)
Ranger Stations	
Coast Guard Stations	CG 79
Pipe Lines	PIPE LINE
Race Tracks or Stadiums	RT
Open-Air Theaters	Open-air theater

Boundaries	International
	State & Provincial
Railroads	Abandoned or Under Construction
	Single Track
	Multiple Track
	Sidings & Spurs
	Overpass
	Underpass
Bridges	Railroad
	Highway
Tunnels	Railroad
	Highway

torque—Any turning or twisting force. Applied to the rolling force imposed on an airplane by the engine in turning the propeller. (*Also see* P-FACTOR.)

touch and go/touch and go landing—An operation by an aircraft that lands and departs on a runway without stopping or exiting the runway.

touchdown RVR—The RVR readout values obtained from RVR equipment serving the runway threshold.

tower communications with, when aircraft transmitter/receiver or both are inoperative—

1. Arrival: (a) receiver inoperative. If you have reason to believe your receiver is inoperative, advise the tower of your type aircraft, position, altitude, and intention to land and request that you be controlled with light signals. (The color, type, and meanings of light signals are published in Federal Aviation Regulation Part 91 and the *Airman's Information Manual,* chapter 4.) When you are approximately 3 to 5 miles from the airport, advise the tower of your position and join the airport traffic pattern. From this point on, watch the tower for light signals. Thereafter, if a complete pattern is made, transmit your position when downwind and/or turning base leg; (b) transmitter inoperative. Join the airport traffic pattern. Monitor the primary local control frequency as depicted on sectional charts for landing or traffic information, and look for a light signal which may be addressed to your aircraft. During hours of daylight, acknowledge tower transmissions or light signals by rocking your wings. At night, acknowledge by blinking the landing or navigation lights; (c) transmitter and receiver inoperative. Join the airport traffic pattern and maintain visual contact with the tower to receive light signals. Acknowledge light signals in accordance with (b) above.

2. Departures: If you experience radio failure prior to leaving the parking area, make every effort to have the equipment repaired. If you are unable to have the malfunction repaired, call the tower by telephone and request authorization to depart without two-way radio communications. If tower authorization is granted, you will be given departure information and requested to monitor the tower frequency or watch for light signals, as appropriate. During daylight hours, acknowledge tower transmissions or light signals by moving the

ailerons or rudder. At night, acknowledge by blinking the landing or navigation lights. If radio malfunction occurs after departing the parking area, watch the tower for light signals or monitor tower frequency.

tower, control—(*See* CONTROL TOWER.)

tower-controlled airports—

1. When operating to an airport where traffic control is being exercised by a control tower, pilots are required to maintain two-way radio contact with the tower while operating within the airport traffic area unless the tower authorizes otherwise. Initial callup should be made about 15 miles from the airport. Unless there is a good reason to leave the tower frequency before exiting the airport traffic area it is a good operating practice to remain on the tower frequency for the purpose of receiving traffic information. In the interest of reducing tower frequency congestion, pilots are reminded that it is not necessary to request permission to leave the tower frequency once outside of the airport traffic area.

2. When necessary, the tower controller will issue clearances or other information for aircraft to generally follow the desired flight path (traffic patterns) when flying in the airport traffic area/control zone, and the proper taxi routes when operating on the ground. If not otherwise authorized or directed by the tower, pilots approaching to land in an airplane must circle the airport to the left, and pilots approaching to land in a helicopter must avoid the flow of fixed wing traffic. However, an appropriate clearance must be received from the tower before landing.

tower en route control (TEC)—TEC is an ATC program to provide a service to aircraft proceeding to and from metropolitan areas. It links designated approach control areas by a network of identified routes made up of the existing airway structure of the National Airspace System. It is an overflow resource in the low altitude system which will enhance ATC services. A few facilities have historically allowed turbojets to proceed between certain city pairs, such as Milwaukee and Chicago, via tower en route and these locations may continue this service. However, the expanded TEC program will be applied, generally, for nonturbojet aircraft operating at and below 10,000 feet. The program is entirely within the approach control airspace of multiple terminal facilities. Essentially, it is for relatively short flights. Participat-

tower en route control (TEC) ***continued***

ing pilots are encouraged to use TEC for flights of two hours duration or less. If longer flights are planned, extensive coordination may be required within the multiple complex which could result in unanticipated delays.

tower radio frequencies—(*See* AERONAUTICAL CHARTS, Figs. A-2 and A-3, and AIRMAN'S INFORMATION MANUAL.)

track—The flight path of an aircraft over the surface of the earth as opposed to intended heading, which may be changed because of wind, etc.

TRACON—Terminal radar approach control.

traffic advisories—Advisories issued to alert a pilot to other known or observed air traffic which may be in such proximity to his aircraft's position or intended route of flight to warrant his attention. Such advisories may be based on:

1. Visual observation from a control tower
2. Observation of radar identified and nonidentified aircraft targets on an ATC radar display, or
3. Verbal reports from pilots or other facilities.

Controllers use the word "traffic" followed by additional information, if known, to provide such advisories, e.g., "Traffic, 2 o'clock, one zero miles, southbound, fast moving, eight thousand."

Traffic advisory service will be provided to the extent possible depending on higher priority duties of the controller or other limitations, e.g., radar limitations, volume of traffic, frequency congestion, or controller workload. Radar/nonradar traffic advisories do not relieve the pilot of his responsibility to see and avoid other aircraft. Pilots are cautioned that there are many times when the controller is not able to give traffic advisories concerning all traffic in the aircraft's proximity; in other words, when a pilot requests or is receiving traffic advisories, he should not assume that all traffic will be issued.

traffic alert—(*See* SAFETY ADVISORY.)

traffic areas, speed in airport—(*See* AIRPORT, OPERATION IN VICINITY OF.)

traffic in sight—Used by pilots to inform a controller that previously issued traffic is in sight.

traffic information (RADAR)—Information issued to alert an aircraft to any radar targets observed on the radar display which may be in such proximity to its position or intended route of flight to warrant its attention.

traffic information service, radar—(*See* RADAR TRAFFIC INFORMATION SERVICE.)

traffic no longer a factor—Indicates that the traffic described in a previously issued traffic advisory is no longer a factor.

traffic pattern—The traffic flow that is prescribed for aircraft landing at, taxiing on, and taking off from an airport. The usual components of a traffic pattern are upwind leg, crosswind leg, downwind leg, base leg, and final approach (see Figure T-3). The following terminology for the various components of a traffic pattern has been adopted as standard for use by control towers and pilots:

Upwind leg—a flight path parallel to the landing runway in the direction of landing

Crosswind leg—a flight path at right angles to the landing runway off its take-off end

Downwind leg—a flight path parallel to the landing runway in the direction opposite to landing

Base leg—a flight path at right angles to the landing runway off its

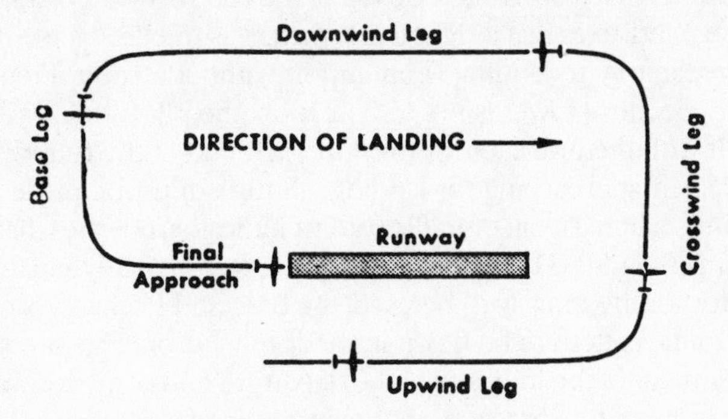

Figure T-3. Traffic pattern.

traffic pattern *continued*

approach end and extending from the downwind leg to the intersection of the extended runway centerline

Final approach—a flight path in the direction of landing along the extended runway centerline from the base leg to the runway.

traffic pattern altitudes—At most airports and military air bases, generally extend from 600 feet to as high as 1,500 feet above the ground for propeller driven aircraft; up to 2,500 feet above the ground for military turbojet aircraft. Therefore, pilots of en route aircraft should be constantly on the alert for other aircraft in traffic patterns and avoid these areas whenever possible. Traffic pattern altitudes should be maintained unless otherwise required by the applicable distance from cloud criteria (Federal Aviation Regulation Part 91).

traffic pattern direction—All airport traffic patterns circle to the left unless approved light signals or visual markings in the segmented circle, or dual runways, indicate a right hand pattern.

traffic pattern indicators—Arranged in pairs in conjunction with landing strip indicators and used to indicate the direction of turns when there is a variation from the normal left traffic pattern. (If there is no segmented circle installed at the airport, traffic pattern indicators may be installed on or near the end of the runway.)

1. Where installed, a flashing amber light near the center of the segmented circle (but usually on top of the control tower or adjoining building) indicates that a right traffic pattern is in effect at the time.

2. Preparatory to landing at an airport without a control tower, or when the control tower is not in operation, the pilot should concern himself with the indicator for the approach end of the runway to be used. When approaching for landing, all turns must be made to the left unless a light signal or traffic pattern indicator indicates that turns should be made to the right. If the pilot will mentally enlarge the indicator for the runway to be used, the base and final approach legs of the traffic pattern to be flown immediately become apparent. Similar treatment of the indicator at the departure end of the runway will clearly indicate the direction of turn after takeoff. (*See* Figure T-4.)

transcribed weather broadcasts—

1. Equipment is provided at selected Federal Aviation Administra-

tion FSSs by which meteorological and notice to airmen data is recorded on tapes and broadcast continuously over the low frequency (200–415kHz) navigational aid (L/MF range or H facility) and VOR.

2. Broadcasts are made from a series of individual tape recordings. The first three tapes identify the station, give general weather forecast conditions in the area, pilot reports (PIREP), radar reports when available, and winds aloft data. The remaining tapes contain weather at selected locations within a 400-mile radius of the central point. Changes, as they occur, are transcribed onto the tapes.

transfer of control—That action whereby the responsibility for the provision of separation to an aircraft is transferred from one controller to another.

transition areas—

1. Controlled airspace extending upward from 700 feet or more above the surface when designated in conjunction with an airport for which an instrument approach procedure has been prescribed, or from 1,200 feet or more above the surface when designated in conjunction with airway route structures or segments. Unless specifically specified otherwise, transition areas terminate at the base of overlying controlled airspace.

2. Transition areas are designated to contain IFR operations in controlled airspace during portions of the terminal operation and while transitioning between the terminal and en route environment.

transmitting in the blind/blind transmission—A transmission from one station to other stations in circumstances where two-way communication cannot be established, but where it is believed that the called stations may be able to receive the transmission.

transponder—The airborne radar beacon receiver–transmitter which automatically receives radio signals from all interrogators on the ground, and which selectively replies with a specific reply pulse or pulse group only to those interrogations being received on the mode to which it is set to respond.

transponder operation—

1. General:

a. Air traffic control radar beacon system (ATCRBS) is similar to and compatible with military coded radar beacon equipment. Civil mode A is identical to military mode 3.

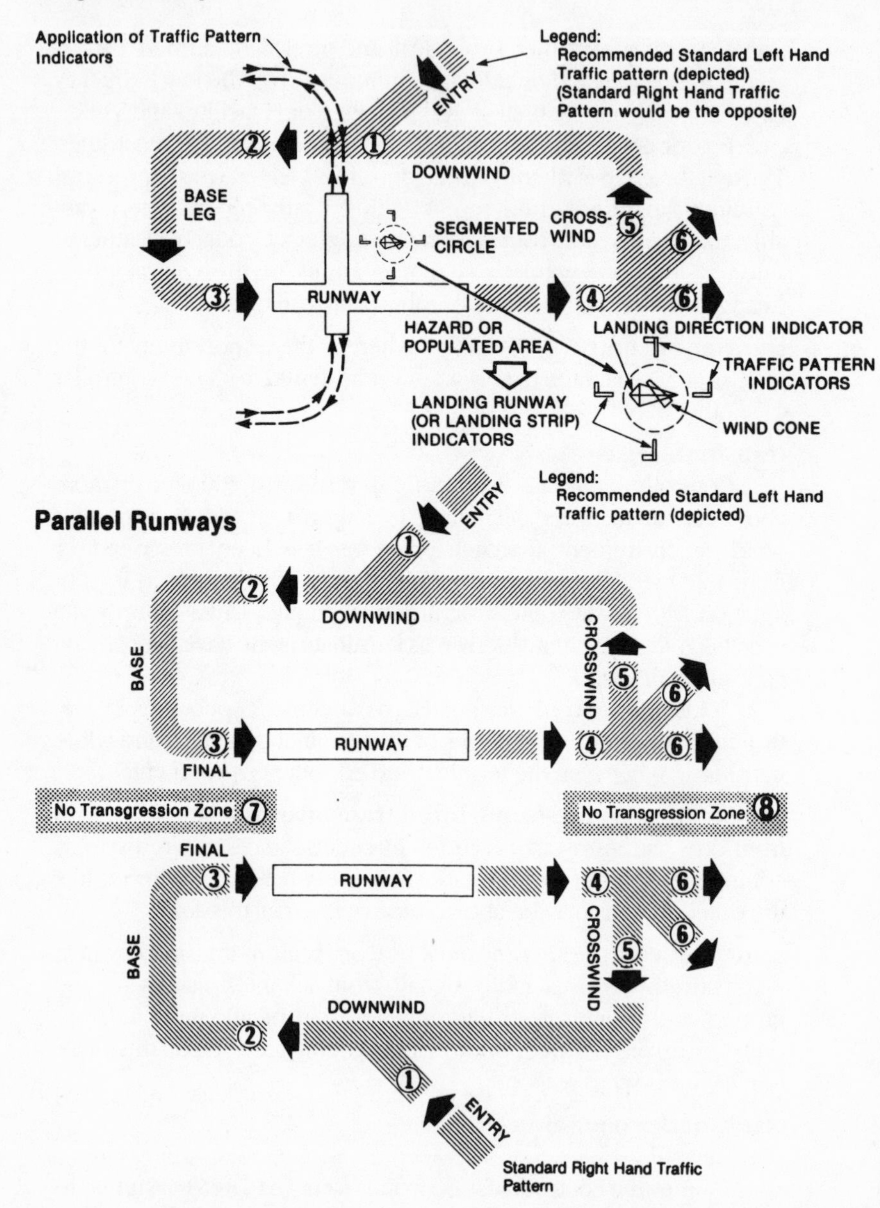

Figure T-4. Application of traffic pattern indicators.

transponder operation *continued*

Key:

1. Enter pattern in level flight, abeam the midpoint of the runway, at pattern altitude. (1000′ AGL is recommended pattern altitude unless established otherwise.)
2. Maintain pattern altitude until abeam approach end of the landing runway, or downwind leg.
3. Complete turn to final at least ¼ mile from the runway.
4. Continue straight ahead until beyond departure end of runway.
5. If remaining in the traffic pattern, commence turn to crosswind leg beyond the departure end of the runway, within 300 feet of pattern altitude.
6. If departing the traffic pattern, continue straight out, or exit with a 45° left turn beyond the departure end of the runway, after reaching pattern altitude.
7. Do not overshoot final or continue on a track which will penetrate the final approach of the parallel runway.
8. Do not continue on a track which will penetrate the departure path of the parallel runway

b. Civil and military transponders should be adjusted to the "on" or normal operating position as late as practicable prior to takeoff and to "off" or "standby" as soon as practicable after completing landing roll, unless the change to "standby" has been accomplished previously at the request of ATC.

c. In all cases, the transponder should be operating while airborne unless otherwise requested by ATC.

d. If entering a United States domestic control area from outside the United States, the pilot should advise on first radio contact with a United States air traffic control center that such equipment is available

transponder operation *continued*

by adding "transponder" or "no code transponder" as appropriate to the aircraft identification.

e. Under no circumstances should a pilot of a civil aircraft operate his transponder on code 0001. This code is reserved for military interceptor operations.

f. It should be noted by all users of the ATC transponders that the coverage they can expect is limited to "line of sight." Low altitudes or aircraft antennae shielding by the aircraft itself may result in reduced range. Range can be improved by climbing to a higher altitude. It may be possible to minimize antenna shielding by locating the antenna where dead spots are only noticed during abnormal flight attitudes.

g. For ATC to utilize one or a combination of the 4096 discrete codes, FOUR DIGIT CODE DESIGNATION will be used, e.g., code 2100 will be expressed as TWO ONE ZERO ZERO.

h. Some transponders are equipped with a mode C automatic altitude reporting capability. This system converts aircraft altitude in 100-foot increments, to coded digital information which is transmitted together with mode C framing pulses to the interrogating radar facility. The manner in which transponder panels are designed differs; therefore a pilot should be thoroughly familiar with the operation of his transponder so that ATC may realize its full capabilities.

2. Visual flight rules (VFR):

a. Adjust transponder to reply on the mode A/3 Code 1200 regardless of altitude, unless otherwise advised by ATC.

b. Adjust transponder to reply on mode C, with altitude reporting capability activated if the aircraft is so equipped, unless deactivation is directed by ATC or unless the installed equipment has not been tested and calibrated as required by Federal Aviation Regulation Part 91. If deactivation is required and your transponder is so designed, turn off the altitude reporting switch and continue to transmit mode C framing pulses. If this capability does not exist, turn off mode C.

3. Emergency operations:

a. When a *distress* or *urgency* condition is encountered, the pilot of an aircraft with a coded radar beacon transponder, who desires to alert a ground radar facility, should squawk MODE 3/A, Code 7700/Emergency and MODE C altitude reporting and then immediately establish communications with the ATC facility.

b. Radar facilities are equipped so that Code 7700 normally triggers an alarm or special indicator at all control positions. Pilots should understand that they might not be within a radar coverage area. Therefore, they should continue squawking Code 7700 and establish radio communications as soon as possible.

NOTE: When making routine code changes, pilots should avoid inadvertent selection of Codes 7500, 7600, or 7700 thereby causing momentary false alarms at automated ground facilities. For example when switching from Code 2700 to Code 7200, switch first to 2200 then to 7200, NOT to 7700 and then 7200. This procedure applies to nondiscrete Code 7500 and all discrete codes in the 7600 and 7700 series (i.e. 7600–7677, 7700–7777) which will trigger special indicators in automated facilities. Only nondiscrete Code 7500 will be decoded as the hijack code.

4. Radio failure: In an aircraft equipped with a coded radar beacon transponder and experiencing a loss of two-way radio capability, the pilot should adjust his transponder to reply on mode A/3, Code 7700 for a period of 1 minute. Then change to Code 7600 for 15 minutes. Repeat steps 1 and 2 as practicable. (*See* SPECIAL EMERGENCY regarding hijacking code.)

5. Radar beacon phraseology: Air traffic controllers, both civil and military, will use the following phraseology when referring to operation of the air traffic control radar beacon system (ATCRBS). Instructions by air traffic control refer only to mode A/3 or mode C operations and do not affect the operation of the transponder on other modes.

SQUAWK (number)—Operate radar beacon transponder on designated code in mode A/3

IDENT—Engage the "IDENT" feature (military I/P) of the transponder

SQUAWK (number) AND IDENT—Operate transponder on specified code in mode A/3 and engage the "IDENT" (military I/P) feature

SQUAWK STANDBY—Switch transponder to standby position

SQUAWK LOW/NORMAL—Operate transponder on low or normal sensitivity as specified. Transponder is operated in "NORMAL" position unless ATC specified "LOW" ("ON" is used instead of "NORMAL" as a master control label on some types of transponders)

SQUAWK ALTITUDE—Activate mode C with automatic altitude reporting

transponder operation *continued*

STOP ALTITUDE SQUAWK—Turn off altitude reporting switch and continue transmitting mode C framing pulses. If your equipment does not have this capability, turn off mode C

STOP SQUAWK (mode in use)—Switch off specified mode. (Used for military aircraft when the controller is unaware if a military service requires the aircraft to continue operating on another mode)

STOP SQUAWK—Switch off transponder

SQUAWK MAYDAY—Operate transponder in the emergency position. (Mode A Code 7700 for civil transponder. Mode 3 Code 7700 and emergency feature for military transponder.)

transponder requirements—Specific details concerning requirements, exceptions and ATC authorized deviations for transponder and mode C operation above 12,500 feet and below 18,000 feet MSL are in FAR Part 91. In general, the FAR requires aircraft to be equipped with mode A/3 (4096 codes) and mode C altitude reporting capability when operating in controlled airspace of the 48 contiguous States and the District of Columbia above 12,500 feet MSL, excluding airspace at and below 2,500 feet AGL. Pilots should insure that their aircraft transponder is operating on an appropriate or ATC assigned VFR/IFR code and mode C when operating in such airspace. If in doubt about the operational status of either feature of your transponder while airborne, contact the nearest ATC facility or FSS and they will advise you what facility you should contact for determining the status of your equipment. Inflight requests for "immediate" deviation may be approved by controllers only when the flight will continue IFR or when weather conditions prevent VFR descent and continued VFR flight in airspace not affected by the FAR. All other requests for deviation should be made by contacting the nearest Flight Service or Air Traffic facility in person or by telephone. The nearest ARTCC will normally be the controlling agency and is responsible for coordinating requests involving deviations in other ARTCC areas.

TRSA—Terminal radar service area.

true airspeed—(*See* AIRSPEED.)

true course—The intended flight path drawn on an aeronautical chart from departure to destination, measured in degrees clockwise from a line projected through true North.

true heading—The true course with a wind-correction angle added or subtracted.

tshwr—Thundershower.

tsmt—Transmit.

tsmtg—Transmitting.

tsmtr—Transmitter.

tstm—Thunderstorm.

turbc—Turbulence.

turn and slip indicator—A gyroscopic instrument for indicating the rate of turning, often combined with a ball bank indicator showing the quality of the turn.

turtleback—The top of the fuselage aft of the cabin; originally detachable in older airplanes.

TV—Television.

twd—Toward.

TWEB—Transcribed weather broadcast.

two-way radio communications failure—(*See* RADIO FAILURE PROCEDURES.)

twr—Tower.

twy—Taxiway.

U

U in the phonetic alphabet is Uniform (you-nee-form).

UFN—Until further notice.

Table U-1.
RADIO PROCEDURES AT UNCONTROLLED AIRPORTS
SUMMARY OF RECOMMENDED COMMUNICATION PROCEDURES

	Facility at Airport	Frequency Use	COMMUNICATION/BROADCAST PROCEDURES	
			Outbound	*Inbound*
1.	UNICOM (No tower or FSS)	Communicate with UNICOM station on published CTAF frequency (122.7, 122.8, or 123.0). If unable to contact UNICOM station, use self-announce procedures on CTAF.	Before taxiing and taking runway for takeoff.	10 miles out— Entering downwind, base, or final. Clear of runway.
2.	No tower, FSS, or UNICOM	Self-announce on CTAF MULTICOM frequency 122.9.	Before taxiing and taking runway for takeoff.	10 miles out— Entering downwind, base, or final. Clear of runway.
3.	No tower, tower closed, FSS open	Communicate with FSS on CTAF frequency.	Before taxiing and taking runway for takeoff.	10 miles out— Entering downwind, base, or final. Clear of runway.

4.	FSS closed* (No tower)	Self-announce on CTAF.	Before taxiing and taking runway for takeoff.	10 miles out—Entering downwind, base, or final. Clear of runway.
5.	Tower closed,* FSS closed, or no FSS	Self-announce on CTAF.	Before taxiing and taking runway for takeoff.	10 miles out—Entering downwind, base, or final. Clear of runway.

* If there is a UNICOM station in operation on the airport, obtain wind and runway information from UNICOM. Return to and monitor CTAF and make self-announce broadcasts as appropriate. The wind direction and runway information may not be available on UNICOM frequency 122.950.

ultimate load—The load which will, or is computed to, cause failure in any structural member.

ultrahigh frequency (UHF)—The frequency band between 300 and 3,000 MHz. The bank of radio frequencies used for military air/ground voice communications. In some instances, this may go as low as 225 MHz and still be referred to as UHF.

ultralight vehicle—A vehicle that:

1. is used or intended to be used for manned operation in the air by a single occupant;

2. is used or intended to be used for recreation or sport purposes only;

3. does not have any U.S. or foreign airworthiness certificate; and

4. if unpowered, weighs less than 155 pounds; or

5. if powered: (a) weighs less than 254 pounds empty weight, excluding floats and safety devices which are intended for deployment in a potentially catastrophic situation; (b) has a fuel capacity not exceeding 5 U.S. gallons; (c) is not capable of more than 55 knots calibrated airspeed at full power in level flight; and (d) has a power-off stall speed which does not exceed 24 knots calibrated airspeed.

Certification and operating rules are covered in Part 103 of the Federal Aviation Regulations.

unable—Radio expression indicates inability to comply with a specific instruction, request, or clearance.

unavbl—Unavailable.

uncontrolled airports—Use left-turn pattern approach unless airport displays approved light signals or visual markings in the segmented circle to indicate right-hand pattern. For recommended radio procedures. (*See* Table U-1.)

uncontrolled airspace—That portion of the airspace that has not been designated as continental control area, control area, control zone, terminal control area, or transition area and within which ATC has neither the authority nor the responsibility for exercising control over air traffic.

unctld—Uncontrolled.

UNICOM—A private aeronautical advisory communications facility operated for purposes other than air traffic control, transmits and receives on one of the following frequencies: 122.7, 122.8, 122.9, 123.0. Locations and frequencies of UNICOMs are shown on aeronautical charts and in the *Airport/Facility Directory.*

unlgtd—Unlighted.

unmkd—Unmarked.

upslp—Upslope.

useful load—In airplanes, the difference, in pounds, between the empty weight and the maximum authorized gross weight.

V

V in the phonetic alphabet is Victor (vik-tah).

variation—Difference between true north and magnetic north.

VASI—Visual approach slope indicator.

vector—The resultant of two quantities (forces, speeds, or deflection) used in aviation to compute load factors, headings, or drift. (*Also see* RADAR VECTOR)

venturi—Or venturi tube, a tube with a restriction used to provide suction to operate flight instruments by allowing the slipstream to pass through it.

verify—Controller expression meaning request confirmation of information, e.g., "verify assigned altitude."

vertical takeoff and landing (VTOL) aircraft—An aircraft which has the capability of vertical take-off and landing. These aircraft include, but are not limited to, helicopters.

vertigo—To earthbound individuals, usally means dizziness or swimming of the head. To a pilot it means, in simple terms, that he doesn't know which end is up. In fact, vertigo during flight can have fatal consequences. On the ground we know which way is up through the combined use of three senses: (1) vision—we can see where we are in relation to fixed objects; (2) pressure—gravitational pull on mus-

vertigo *continued*

cles and joints tells us which way is down; and (3) special parts in our inner ear—the otoliths—tell us which way is down by gravitational pull. It should be noted that accelerations of the body are detected by the fluid in the semicircular canals of the inner ear, and this tells us when we change position. However, in the absence of a visual reference, such as flying into a cloud or overcast, the accelerations can be confusing, especially since their forces can be misinterpreted as gravitational pulls on the muscles and otoliths. The result is often disorientation and vertigo (or dizziness). All pilots should have an instructor pilot produce maneuvers which will produce the sensation of vertigo. Once experienced, later unanticipated incidents of vertigo can be overcome. Closing the eyes for a second or two may help, as will watching the flight instruments, believing them, and controlling the airplane in accordance with the information presented on the instruments. All pilots should obtain the minimum training recommended by the Federal Aviation Administration for attitude control of aircraft solely by reference to the gyroscopic instruments. Pilots are susceptible to experiencing vertigo at night, and in any flight condition when outside visibility is reduced to the point that the horizon is obscured. An additional type of vertigo is known as flicker vertigo. Light, flickering at certain frequencies, from 4 to 20 times per second, can produce unpleasant and dangerous reactions in some persons, including nausea, dizziness, unconsciousness, or even reactions similar to epileptic fit. In a single engine propeller airplane heading into the sun, the propeller may cut the sun to give this flashing effect, particularly during landings when the engine is throttled back. These undesirable effects may be avoided by not staring directly through the prop for more than a moment, and by making frequent but small changes in RPM. The flickering light traversing helicopter blades has been known to cause this difficulty, as has the bounce-back from rotating beacons on aircraft which have penetrated clouds. For this reason pilots should turn off rotating beacons or high intensity aircraft strobe lights when penetrating clouds.

very high frequency (VHF)—The frequency band between 30 and 300 MHz. Portions of this band, 108 to 118 MHz, are used for certain NAVAIDS; 118 to 136 MHz are used for civil air/ground voice communications. Other frequencies in this band are used for purposes not related to air traffic control.

very low frequency (VLF)—The frequency band between 3 and 30 kHz.

very pistol—A piece of emergency equipment for firing colored flare signals at night; more common expression is pyrotechnic signal device or flare gun.

VFR—Visual flight rules.

VFR advisory service—

1. Is provided by numerous nonradar approach control facilities to those pilots intending to land at an airport served by an approach control tower. This service includes: wind, runway, traffic, and NOTAM information, unless this information is contained in the ATIS broadcast and the pilot indicates he has received the ATIS information.
2. Such information will be furnished upon initial contact with concerned approach control facility. The pilot will be requested to change to the tower frequency at a predetermined time or point, to receive further landing information.
3. Where available, use of this procedure will not hinder the operation of VFR flights by requiring excessive spacing between aircraft or devious routing. Radio contact points will be based on time or distance rather than on landmarks.
4. Compliance with this procedure is not mandatory but pilot participation is encouraged.

VFR at night—In addition to the altitude appropriate for the direction of flight, pilots should maintain an altitude which is at or above the minimum en route altitude as shown on charts. This is especially true in mountainous terrain, where there is usually very little ground reference. Don't depend on your being able to see those built-up rocks or TV towers in time to miss them.

VFR clearances, special—

1. An ATC clearance must be obtained prior to operating within a control zone when the weather is less than that required for VFR flight. Within most control zones, a VFR pilot may request and be given a clearance to conduct special VFR flight to, from, or within the

VFR clearances, special *continued*

control zone providing such flight will not delay IFR operations. The weather and clearance from cloud requirements for special VFR flight are: 1 mile ground visibility if landing or departing, 1 mile flight visibility if transiting the control zone, and flight to be conducted clear of clouds. When a control tower is located within the control zone, requests for clearances should be to the tower. If no tower is located within the control zone, a clearance may be obtained from the nearest tower, flight service station, or center.

2. It is not necessary to file a complete flight plan with the request for clearance, but the pilot should state his intentions in sufficient detail to permit air traffic control to fit his flight into the traffic flow. The clearance will not contain a specific altitude, as the pilot must remain clear of clouds. The controller may require the pilot to fly at or below a certain altitude due to other traffic, but the altitude specified will permit flight at or above the minimum safe altitude. In addition, at radar locations flights may be vectored if necessary for control purposes or on pilot request.

3. ATC provides separation between special VFR flights and between them and other IFR flights.

4. Within some control zones, the volume of IFR traffic is such that special VFR flight cannot be permitted. A list of these control zones is in FAR Part 93 and on sectional aeronautical charts.

Special VFR clearances are effective within control zones only. ATC does not provide separation after an aircraft leaves the control zone on a special VFR clearance.

Special VFR operations by fixed-wing aircraft are prohibited between sunset and sunrise unless the pilot is instrument rated and the aircraft is equipped for IFR flight.

VFR conditions—Basic weather conditions prescribed for flight under visual flight rules. (*See* VFR FLIGHT WEATHER MINIMUMS)

VFR conditions, special (special VFR minimum weather conditions)—Weather conditions which are less than basic VFR weather conditions and which permit flight by some aircraft under visual flight rules.

UNDER VFR-More than 3,000' above the surface.

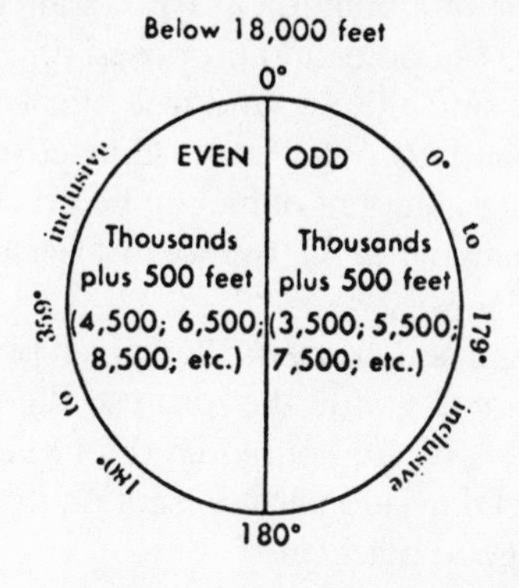

COURSES ARE MAGNETIC

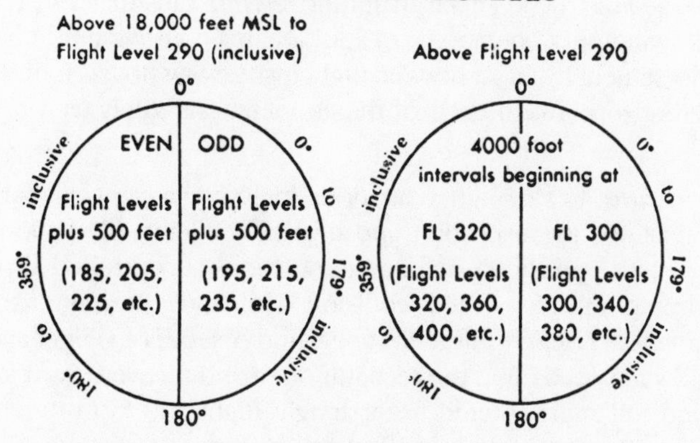

Figure V-1. VFR cruising altitudes/flight levels.

VFR cruising altitudes/flight levels—Controlled and uncontrolled airspace (*See* Figure V-1).

VFR flight plan, compulsory—VFR flights into a coastal or domestic ADIZ/DEWIZ are required to file a VFR flight plan for security purposes. ADIZ procedures are found in the National Security section of Federal Aviation Regulations.

VFR flight plan—Unless otherwise authorized by ATC, each person filing a VFR flight plan shall include in it the following information: (1) aircraft identification number; (2) type of aircraft; (3) full name and address of pilot; (4) point and proposed time of departure; (5) proposed route, cruising altitude, and true airspeed at that altitude; (6) first intended landing point and estimated elapsed time; (7) alternate airport(s); (8) amount of fuel on board (in hours); and (9) in case of an international flight, number of persons in the aircraft. (*See* Figures V-2 and V-3.)

To obtain maximum benefits of the flight plan program, flight plans should be filed directly with the nearest flight service station. Although position reports are not required for VFR flight plans, periodic reports to Federal Aviation Administration flight service stations along the route are good practice.

VFR flight proficiency—The applicant for a private pilot certificate must have logged instruction from an authorized flight instructor in at least the following pilot operations. In addition, his logbook must contain an endorsement by an authorized flight instructor who has found him competent to perform each of those operations safely as a private pilot.

In airplanes. (1) Preflight operations, including weight and balance determination, line inspection, and airplane servicing; (2) Airport and traffic pattern operations, including operations at controlled airports, radio communications, and collision avoidance precautions; (3) Flight maneuvering by reference to ground objects; (4) Flight at critically slow airspeeds, and the recognition of and recovery from imminent and full stalls entered from straight flight and from turns; (5) Normal and crosswind takeoffs and landings; (6) Control and maneuvering an airplane solely by reference to instruments, including descents and climbs using radio aids or radar directives; (7) Cross-country flying, using pilotage, dead reckoning, and radio aids, including one 2-hour flight; (8) Maximum performance takeoffs and landings; (9) Night flying, including takeoffs, landings, and VFR navigation; and (10) Emergency operations, including simulated aircraft and equipment malfunctions.

Form Approved: OMB No. 2120-0026

U.S. DEPARTMENT OF TRANSPORTATION FEDERAL AVIATION ADMINISTRATION **FLIGHT PLAN**	(FAA USE ONLY) ☐ PILOT BRIEFING ☐ VNR ☐ STOPOVER	TIME STARTED	SPECIALIST INITIALS

1. TYPE VFR IFR DVFR	2. AIRCRAFT IDENTIFICATION	3. AIRCRAFT TYPE/ SPECIAL EQUIPMENT	4. TRUE AIRSPEED KTS	5. DEPARTURE POINT	6. DEPARTURE TIME PROPOSED (Z) / ACTUAL (Z)	7. CRUISING ALTITUDE

8. ROUTE OF FLIGHT

9. DESTINATION (Name of airport and city)	10. EST. TIME ENROUTE HOURS / MINUTES	11. REMARKS

12. FUEL ON BOARD HOURS / MINUTES	13. ALTERNATE AIRPORT(S)	14. PILOT'S NAME, ADDRESS & TELEPHONE NUMBER & AIRCRAFT HOME BASE 17. DESTINATION CONTACT/TELEPHONE (OPTIONAL)	15. NUMBER ABOARD

16. COLOR OF AIRCRAFT	CIVIL AIRCRAFT PILOTS. FAR Part 91 requires you file an IFR flight plan to operate under instrument flight rules in controlled airspace. Failure to file could result in a civil penalty not to exceed $1,000 for each violation (Section 901 of the Federal Aviation Act of 1958, as amended). Filing of a VFR flight plan is recommended as a good operating practice. See also Part 99 for requirements concerning DVFR flight plans.

FAA Form 7233-1 (8-82) CLOSE VFR FLIGHT PLAN WITH ______________ FSS ON ARRIVAL

Figure V-2. VFR flight plan.

SCALE 1:1,000,000 WORLD AERONAUTICAL CHARTS

PILOT'S PREFLIGHT CHECK LIST

DATE

WEATHER ADVISORIES	ALTERNATE WEATHER	NOTAMS
EN ROUTE WEATHER	FORECASTS	AIRSPACE RESTRICTIONS
DESTINATION WEATHER	WINDS ALOFT	MAPS

FLIGHT LOG

DEPARTURE POINT	VOR IDENT. / FREQ.	RADIAL TO / FROM	DISTANCE LEG / REMAINING	TIME PT-TO-PT / CUMULATIVE	TIME TAKEOFF	GROUND SPEED
CHECK POINT					ETA / ATA	
DESTINATION			TOTAL			

POSITION REPORT: FVFR report hourly, IFR as required by ATC

ACFT. IDENT.	POSITION	TIME	ALT.	IFR/VFR	EST. NEXT FIX	NAME OF SUCCEEDING FIX	PIREPS

REPORT CONDITIONS ALOFT—
CLOUD TOPS, BASES, LAYERS, VISIBILITY, TURBULENCE, HAZE, ICE, THUNDERSTORMS

CLOSE FLIGHT PLAN UPON ARRIVAL

SECTIONAL AERONAUTICAL CHARTS SCALE 1:500,000

C48—16—79530-1 GPO . 1967 O—261-206

Figure V-3. VFR flight plan, pilot's preflight check list.

VFR flight weather minimums in controlled airspace—

MINIMUM VISIBILITY AND DISTANCE FROM CLOUDS (VFR)

Altitude	*Flight Visibility*	*Distance from Clouds*
1,200 feet or less above the surface (regardless of MSL altitude)	3 statute miles	500 feet below 1,000 feet above 2,000 feet horizontal
More than 1,200 feet above the surface but less than 10,000 feet MSL	3 statute miles	500 feet below 1,000 feet above 2,000 feet horizontal
More than 1,200 feet above the surface and at or above 10,000 feet MSL	5 statute miles	1,000 feet above 1,000 feet below 1 mile horizontal

In addition to the above, when operating within a control zone beneath a ceiling, the ceiling must not be less than 1,000 feet. If the pilot intends to land or take off or enter a traffic pattern within a control zone, the ground visibility must be at least 3 miles at that airport. If ground visibility is not reported at the airport, 3 miles flight visibility is required (Federal Aviation Regulation Part 91).

VFR flight weather minimums in uncontrolled airspace—

MINIMUM VISIBILITY AND DISTANCE FROM CLOUDS (VFR)

Altitude	*Flight Visibility*	*Distance from Clouds*
1,200 feet or less above the surface (regardless of MSL altitude)	1 statute mile	Clear of clouds
More than 1,200 feet above the surface but less than 10,000 feet MSL	1 statute mile	500 feet below 1,000 feet above 2,000 feet horizontal
More than 1,200 feet above the surface and at or above 10,000 feet MSL	5 statute miles	1,000 feet below 1,000 feet above 1 mile horizontal

VFR in a congested area?—Listen! As pointed out in the KEEP-EM-HIGH program: "A high percentage of near midair collisions occur below 8,000 feet AGL and within 30 miles of an airport. . . ." When operating VFR in these highly congested areas, whether you intend to

VFR in a congested area? ***continued***

land at an airport within the area or are just flying through, it is recommended that extra vigilance be maintained and that you monitor an appropriate control frequency. Normally the appropriate frequency is an approach control frequency. By such monitoring action you can "get the picture" of the traffic in your area. When the approach controller has radar, traffic advisories may be given to VFR pilots who request them, subject to the provisions included in radar traffic information service.

VFR military training routes (VR)—Routes used by the Department of Defense and associated Reserve and Air Guard units for the purpose of conducting low-altitude navigation and tactical training under VFR below 10,000 feet MSL at airspeeds in excess of 250 knots IAS. They are shown as numbered gray lines (e.g., VR 1007) on sectional charts.

VFR not recommended—An advisory provided by a flight service station to a pilot during a preflight or inflight weather briefing that flight under visual flight rules is not recommended. To be given when the current and/or forecasted weather conditions are at or below VFR minimums. It does not abrogate the pilot's authority to make his own decision.

VFR radar assistance—

1. Radar equipped FAA ATC facilities provide radar assistance and navigation service (vectors) to VFR aircraft provided the aircraft can communicate with the facility, are within radar coverage, and can be radar identified.

2. Pilots should clearly understand that authorization to proceed in accordance with such radar navigational assistance does not constitute authorization for the pilot to violate FARs. In effect, assistance provided is on the basis that navigational guidance information issued is advisory in nature and the job of flying the aircraft safely remains with the pilot.

3. In many cases, the controller will be unable to determine if flight into instrument conditions will result from his instructions. To avoid possible hazards resulting from being vectored into IFR conditions,

pilots should keep the controller advised of the weather conditions in which he is operating and along the course ahead.

4. Radar navigation assistance (vectors) may be initiated by the controller when one of the following conditions exist:

a. The controller suggests the vector and the pilot concurs.

b. A special program has been established and vectoring service has been advertised.

c. In the controller's judgment the vector is necessary for air safety.

5. Radar navigation assistance (vectors) and other radar derived information may be provided in response to pilot requests. Many factors, such as limitations of radar, volume of traffic, communications frequency, congestion, and controller workload could prevent the controller from providing it. The controller has complete discretion for determining if he is able to provide the service in a particular case. His decision not to provide the service in a particular case is not subject to question.

VFR report—On a VFR flight, when making initial radio contact with a control tower or a flight service station providing airport advisory service, report position and altitude. This will make it easier for ground facility personnel and other pilots to locate your aircraft.

VFR requirements—(*See* VFR FLIGHT WEATHER MINIMUMS IN CONTROLLED AIRSPACE; VFR FLIGHT WEATHER MINIMUMS IN UNCONTROLLED AIRSPACE.)

VFR, special—(*See* SPECIAL VFR CLEARANCES, SPECIAL VFR CONDITIONS, and SPECIAL VFR OPERATIONS.)

VFR operations, special—Aircraft operating in accordance with clearances within control zones in weather conditions less than the basic VFR weather minima.

VHF omnidirectional range (VOR)—

1. A radio navigation aid, VORs operate within the 108.0–117.95 MHz frequency band and have a power output necessary to provide coverage within their assigned operational service volume. The equipment is VHF; thus, it is subject to line-of-sight restriction, and its range varies proportionally to the altitude of the receiving equipment. There is some "spillover," however, and reception at an altitude of

VHF omnidirectional range (VOR) ***continued***

1,000 feet is about 40 to 45 miles. This distance increases with altitude.

2. There is voice transmission on the VOR frequency and all information broadcast over L/MF ranges is also available over the VORs.

3. The effectiveness of the VOR depends upon proper use and adjustment of both ground and airborne equipment: (a) accuracy—the accuracy of course alignment of the VOR is excellent, being generally plus or minus 1 degree; (b) roughness—on some VORs, minor course roughness may be observed, evidenced by course needle or brief flag alarm activity (some receivers are more subject to these irregularities than others). At a few stations, usually in mountainous terrain, the pilot may occasionally observe a brief course needle oscillation, similar to the indication of "approaching station." Pilots flying over unfamiliar routes are cautioned to be on the alert for these vagaries and, in particular, to use the "to-from" indicator to determine positive station passage; (c) certain propeller RPM settings can cause the VOR course deviation indicator to fluctuate as much as ±6 degrees. Slight changes to the RPM setting will normally smooth out this roughness. Helicopter rotor speeds may also cause VOR course disturbances. Pilots are urged to check for this propeller modulation phenomenon prior to reporting a VOR station or aircraft equipment for unsatisfactory operation.

4. The only positive method of identifying a VOR is by its Morse code identification or by the recorded automatic voice identification, which is always indicated by use of the word "VOR" following the name of the range. Reliance on determining the identification of an omnirange should never be placed on listening to voice transmissions by the flight service station (or approach control facility) involved. Many FSSs remotely operate several omniranges which have different names from each other and in some cases none has the name of the parent FSS. (During periods of maintenance the coded identification is removed.)

5. Voice identification has been added to numerous VHF omniranges. The transmission consists of a voice announcement, "AIRVILLE VOR" (VORTAC) alternating with the usual Morse code identi-

fication. If no air-ground communications facility is associated with the omnirange, "AIRVILLE UNATTENDED VOR" (VORTAC) will be heard.

VHF omnidirectional range/tactical air navigation (VORTAC)—

1. VORTAC is a facility consisting of two components, VOR and TACAN, which provides three individual services: VOR azimuth, TACAN azimuth, and TACAN distance (DME) at one site. Although consisting of more than one component, incorporating more than one operating frequency, and using more than one antenna system, a VORTAC is considered to be a unified navigational aid. Both components of a VORTAC are envisioned as operating simultaneously and providing the three services at all times.

2. Transmitted signals of VOR and TACAN are each identified by three-letter code transmission and are interlocked so that pilots using VOR azimuth with TACAN distance can be assured that both signals being received are definitely from the same ground station. The frequency channels of the VOR and the TACAN at each VORTAC facility are "paired" in accordance with a national plan to simplify airborne operation.

vicinity of an airport, flight in—(*See* OPERATIONS and NONTOWER AIRPORTS.)

victor airways—Low altitude federal airways that are preestablished flight routes between VOR omnidirectional radio stations as shown on sectional charts. East–west airways are even-numbered; north–south, odd-numbered.

viscosity—The measure of body, or "thickness" in a fluid. Important in determining the correct lubricating oil for any engine.

visibility—The ability, as determined by atmospheric conditions and expressed in units of distance, to see and identify prominent unlighted objects by day and prominent lighted objects by night. Visibility is reported as statute miles, hundreds of feet, or meters. (Refer to FAR Part 91.)

1. Flight Visibility—The average forward horizontal distance, from the cockpit of an aircraft in flight, at which prominent unlighted objects may be seen and identified by day and prominent lighted objects may be seen and identified by night.

visibility *continued*

2. Ground Visibility—Prevailing horizontal visibility near the earth's surface as reported by the United States National Weather Service or an accredited observer.

3. Prevailing Visibility—The greatest horizontal visibility equaled or exceeded throughout at least half the horizon circle which need not necessarily be continuous.

4. Runway Visibility Value (RVV)—The visibility determined for a particular runway by a transmissometer. A meter provides a continuous indication of the visibility (reported in miles or fraction of miles) for the runway. RVV is used in lieu of prevailing visibility in determining minimums for a particular runway.

5. Runway Visual Range (RVR)—An instrumentally derived value, based on standard calibrations, that represents the horizontal distance a pilot will see down the runway from the approach end. It is based on the sighting of either high intensity runway lights or on the visual contrast of other targets, whichever yields the greater visual range. RVR, in contrast to prevailing or runway visibility, is based on what a pilot in a moving aircraft should see looking down the runway. RVR is horizontal visual range, not slant visual range. It is based on the measurement of a transmissometer made near the touchdown point of the instrument runway and is reported in hundreds of feet. RVR is used in lieu of RVV and/or prevailing visibility in determining minimums for a particular runway.

a. Touchdown RVR—The RVR visibility readout values obtained from RVR equipment serving the runway touchdown zone.

b. Mid-RVR—The RVR readout values obtained from RVR equipment located midfield of the runway.

c. Rollout RVR—The RVR readout values obtained from RVR equipment located nearest the rollout end of the runway.

visibility minimums—(*See* VFR FLIGHT WEATHER MINIMUMS IN CONTROLLED AIRSPACE; VFR FLIGHT WEATHER MINIMUMS IN UNCONTROLLED AIRSPACE.)

vision—On the ground, reduced or impaired vision can sometimes be dangerous depending on where you are and what you are doing.

In flying it is always dangerous. On the ground or in the air, a number of factors such as hypoxia, carbon monoxide, alcohol, drugs, fatigue, or even bright sunlight can affect your vision. In the air these effects are critical. Some good specific rules are: make sure of sunglasses on bright days to avoid eye fatigue; during night flights use red covers on the flashlights to avoid destroying any dark adaption; remember that drugs, alcohol, heavy smoking, and the other factors mentioned above have early effects on visual acuity.

visual approach—An approach wherein an aircraft on an IFR flight plan, operating in VFR conditions under the control of a radar facility and having an air traffic control authorization, may deviate from the prescribed instrument approach procedure and proceed to the airport of destination by visual reference to the surface.

visual approach slope indicator (VASI)—

1. The VASI is a system of lights so arranged to provide visual descent guidance information during the approach to a runway. These lights are visible from 3–5 miles during the day and up to 20 miles or more at night. The visual glide path of the VASI provides safe obstruction clearance within plus or minus 10 degrees of the extended runway centerline and to 4 NM from the runway threshold. Descent, using the VASI, should not be initiated until the aircraft is visually aligned with the runway. Lateral course guidance is provided by the runway or runway lights.

2. VASI installations may consist of either 2, 4, 6, 12, or 16 light units arranged in bars referred to as near, middle, and far bars. Most VASI installations consist of 2 bars, near and far, and may consist of 2, 4, or 12 light units. Some airports have VASIs consisting of three bars, near, middle, and far, which provide an additional visual glide path for use by high cockpit aircraft. This installation may consist of either 6 or 16 light units. VASI installations consisting of 2, 4, or 6 light units are located on one side of the runway, usually the left. Where the installation consists of 12 or 16 light units, the light units are located on both sides of the runway.

3. Two-bar VASI installations provide one visual glide path which is normally set at 3 degrees. Three-bar VASI installations provide two visual glide paths. The lower glide path is provided by the near and

visual approach slope indicator (VASI) ***continued***

middle bars and is normally set at 3 degrees while the upper glide path, provided by the middle and far bars, is normally 1/4 degree higher. This higher glide path is intended for use only by high cockpit aircraft to provide a sufficient threshold crossing height. Although normal glide path angles are three degrees, angles at some locations may be as high as 4.5 degrees to give proper obstacle clearance. Pilots of high performance aircraft are cautioned that use of VASI angles in excess of 3.5 degrees may cause an increase in runway length required for landing and rollout.

4. The following information is provided for pilots as yet unfamiliar with the principles and operation of this system and pilot technique required. The basic principle of the VASI is that of color differentiation between red and white. Each light unit projects a beam of light having a white segment in the upper part of the beam and red segment in the lower part of the beam. The light units are arranged so that the pilot using the VASIs during an approach will see the combination of lights shown below.

5. 2-bar VASI (4 light units shown)

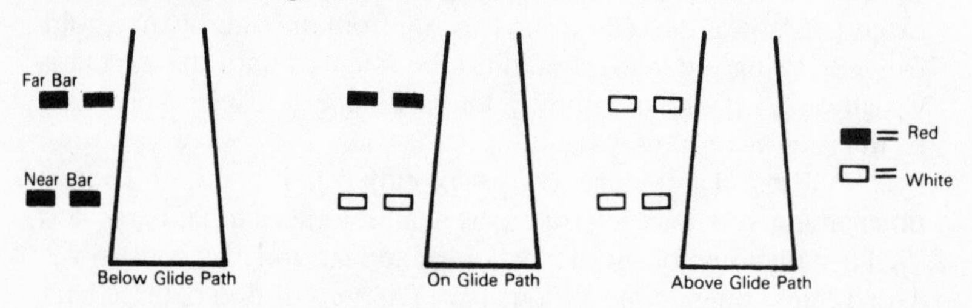

6. 3-bar VASI (6 light units shown)

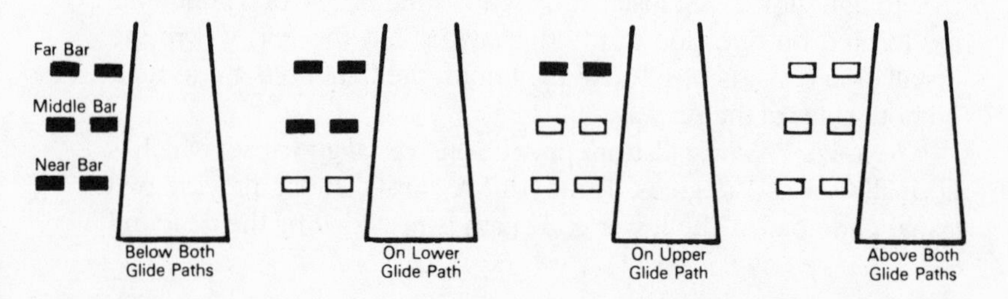

7. Other VASI configurations

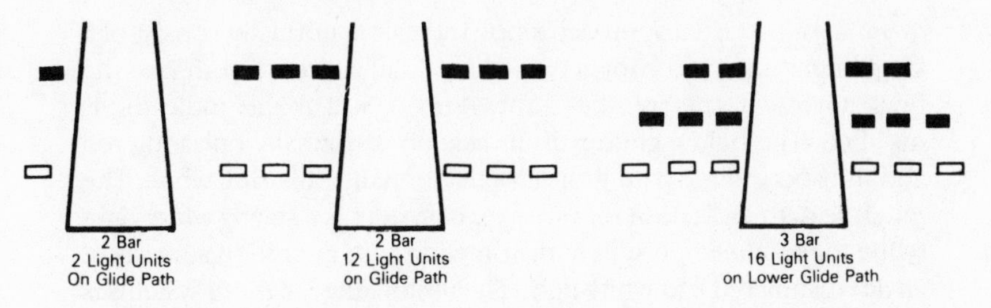

visual approach slope systems—

1. Several types of nonstandard visual approach slope indicators have been installed at general aviation and air carrier airports.

2. Tricolor visual approach slope indicators normally consist of a single light unit projecting a three-color visual approach path into the final approach area of the runway upon which the indicator is installed. The below glide path indication is red, the above glide path indication is amber, and the on glide path indication is green. These types of indicators have a useful range of approximately one-half to one mile during the day and up to five miles at night depending upon the visibility conditions.

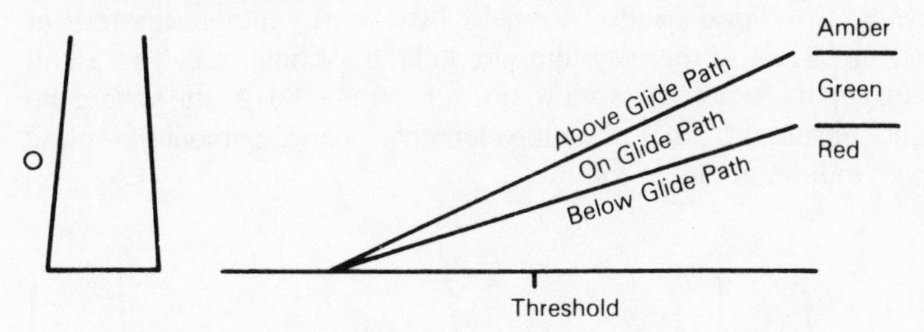

CAUTION: Pilots should be aware that this yellow-green-red configuration produces a yellow-green transition light beam between the yellow and green primary light segments and an anomalous yellow transition light beam between the green and red primary light segments. This anomalous yellow signal could cause confusion with the primary yellow too-high signal.

visual approach slope systems *continued*

3. Pulsating visual approach slope indicators normally consist of a single light unit projecting a two-color visual approach path into the final approach area of the runway upon which the indicator is installed. The below glide path indication is normally pulsating red and the above glide path indication is normally pulsating white. The on glide path indication for one type of system is a steady white light while for another type system the on glide path indication consists of an alternating red and white light. The useful range of these systems is about four miles during the day and up to ten miles at night.

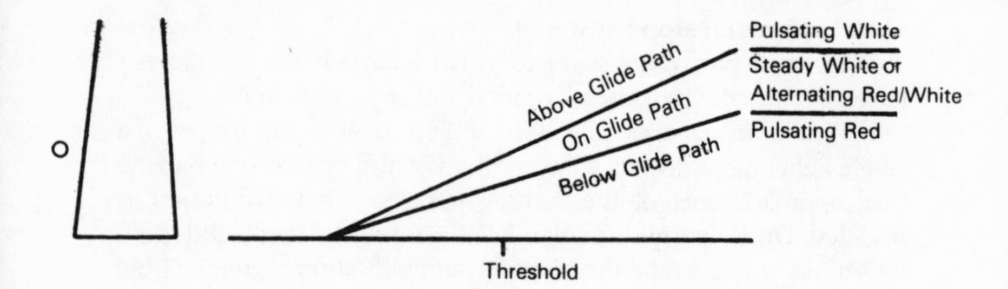

4. Alignment of elements systems are installed on some small general aviation airports and are a low-cost system consisting of painted plywood panels, normally black and white or fluorescent orange. Some of these systems are lighted for night use. The useful range of these systems is approximately ¾ mile. To use the system the pilot positions his aircraft so the elements are in alignment. The glide path indications are as follows.

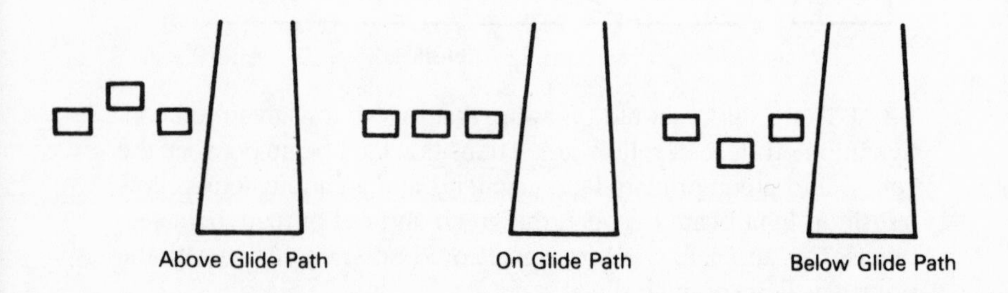

5. The precision approach path indicator (PAPI) uses light units similar to the VASI but are installed in a single row of either two or four light units. These systems have an effective visual range of about 5 miles during the day and up to 20 miles at night. The row of light units is normally installed on the left side of the runway and the glide path indications are as follows.

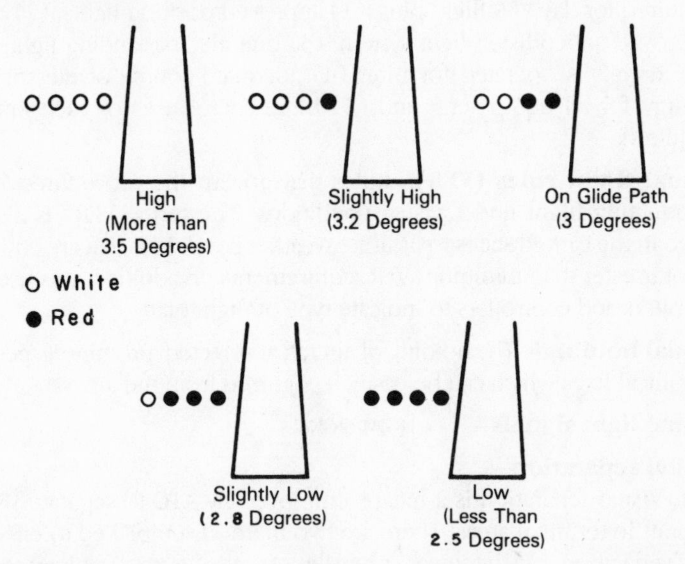

visual codes, ground-air for use by ground search parties and survivors—(*See* EMERGENCY VISUAL CODES.)

visual emergency signals—(*See* Figure V-4.)

visual flight rules, required equipment, day—The following instruments and equipment are required: (1) airspeed indicator; (2) altimeter; (3) magnetic direction indicator; (4) tachometer for each engine; (5) oil pressure gauge for each engine using pressure system; (6) temperature gauge for each liquid-cooled engine; (7) oil temperature gauge for each air-cooled engine; (8) manifold pressure gauge for each altitude engine; (9) fuel gauge for each tank; (10) landing gear position indicator, if the aircraft has retractable landing gear; (11) at least one pyrotechnic signalling device, and approved flotation gear readily available to each occupant, if the aircraft is operated for hire

visual flight rules, required equipment, day *continued*

over water and beyond power-off gliding distance from shore; (12) approved safety belts and shoulder harnesses where appropriate for all occupants.

visual flight rules, required equipment, night—All equipment required for day VFR flight plus: (1) approved position lights; (2) an approved anticollision light system; (3) one electric landing light, if the aircraft is operated for hire; (4) adequate source of electrical energy for all equipment; and (5) three spare fuses of each kind required.

visual flight rules (VFR)—Rules that govern the procedures for conducting flight under visual conditions. The term "VFR" is also used in the United States to indicate weather conditions that are equal to or greater than minimum VFR requirements. In addition, it is used by pilots and controllers to indicate type of flight plan.

visual holding—The holding of aircraft at selected, prominent, geographical fixes which can be easily recognized from the air.

visual light signals—(*See* LIGHT SIGNALS.)

visual separation—

1. Visual separation is a means employed by ATC to separate IFR aircraft in terminal areas. There are two methods employed to effect this separation: (a) the tower controller sees the aircraft involved and issues instructions, as necessary, to ensure that the aircraft avoid each other; (b) a pilot sees the other aircraft involved and upon instructions from the controller provides his own separation by maneuvering his aircraft as necessary to avoid it. This may involve following in-trail behind another aircraft or keeping it in sight until it is no longer a factor.

2. When pilots have been told to follow another aircraft or to provide visual separation from it, they should promptly notify the controller if they do not sight the other aircraft involved.

3. A pilot's acceptance of instructions to follow another aircraft or provide visual separation from it is an acknowledgment that the pilot will maneuver his aircraft as necessary to avoid the other aircraft or to

EMERGENCY SERVICES AVAILABLE TO PILOTS

VISUAL EMERGENCY SIGNALS

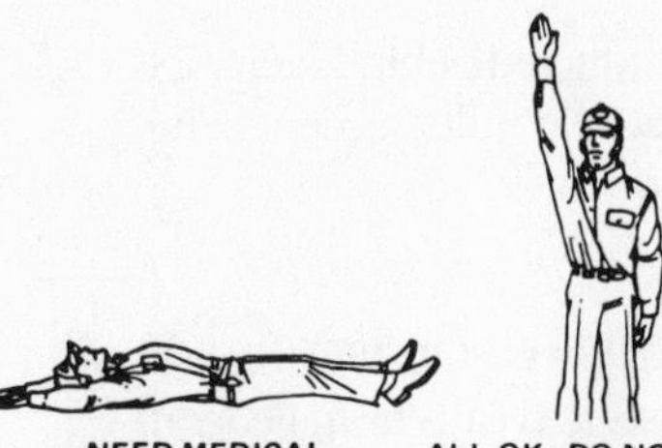

NEED MEDICAL ASSISTANCE–URGENT
Used only when life is at stake

ALL OK–DO NOT WAIT
Wave one arm overhead

CAN PROCEED SHORTLY–WAIT IF PRACTICABLE
One arm horizontal

NEED MECHANICAL HELP OR PARTS–LONG DELAY
Both arms horizontal

USE DROP MESSAGE
Make throwing motion

OUR RECEIVER IS OPERATING
Cup hands over ears

DO NOT ATTEMPT TO LAND HERE
Both arms waved across face

LAND HERE
Both arms forward horizontally, squatting and point in direction of landing–Repeat

NEGATIVE (NO)
White cloth waved horizontally

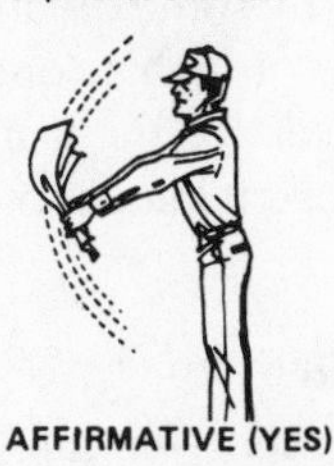

AFFIRMATIVE (YES)
White cloth waved vertically

PICK US UP–PLANE ABANDONED
Both arms vertical

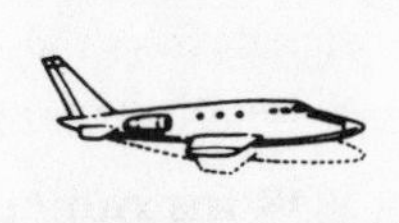

AFFIRMATIVE (YES)
Dip nose of plane several times

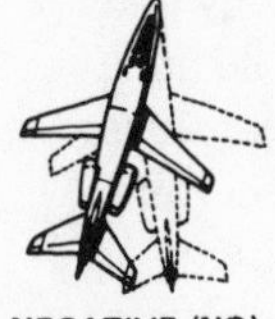

NEGATIVE (NO)
Fishtail plane

HOW TO USE THEM

If you are forced down and are able to attract the attention of the pilot of a rescue airplane, the body signals illustrated on this page can be used to transmit messages to him as he circles over your location. Stand in the open when you make the signals. Be sure that the background, as seen from the air is not confusing. Go through the motions slowly and repeat each signal until you are positive that the pilot understands you.

Figure V-4. Visual emergency signals.

visual separation *continued*

maintain in-trail separation. In operations conducted behind heavy jet aircraft, it is also an acknowledgment that the pilot accepts the responsibility for wake turbulence separation.

VOR—VHF omnidirectional range.

VOR communications guard on VFR flights—On VFR flights, guard the voice channel of VORs for broadcasts and calls from Federal Aviation Administration flight service stations. Where the VOR voice channel is being utilized for ATIS broadcasts, pilots of VFR flights are urged to guard the voice channel of an adjacent VOR. When in contact with a control facility, notify the controller if you plan to leave the frequency. This could save the controller time by not trying to call you on that frequency.

VOR equipment—Until planned receiving equipment can be installed at remote VOR transmitting sites, pilots should remember that while they may be hearing the station loud and clear through the voice facility of the VOR to which they are tuned, the receivers being used by the station they are calling may be some distance away, rather than at the transmitter site. In the case of flights conducted at fairly low altitudes, this may make it difficult for the station to receive signals from the aircraft and should be taken into account when trying to establish two-way communication.

VOR instrument examples—(*See* Figure V-5.)

VOR receiver check—Periodic VOR receiver calibration is most important. If a receiver's automatic gain control or modulation circuit deteriorates, it is possible for it to display acceptable accuracy and sensitivity close in to the VOR or VOT and display out-of-tolerance readings when located at greater distances where weaker signal areas exist. The likelihood of this deterioration varies between receivers, and is generally considered a function of time. The best assurance of having an accurate receiver is periodic calibration. Yearly intervals are recommended at which time an authorized repair facility should recalibrate the receiver to the manufacturer's specifications.

VOR test signal (VOT)—A ground facility which emits a test signal to check VOR receiver accuracy. System is limited to ground use only.

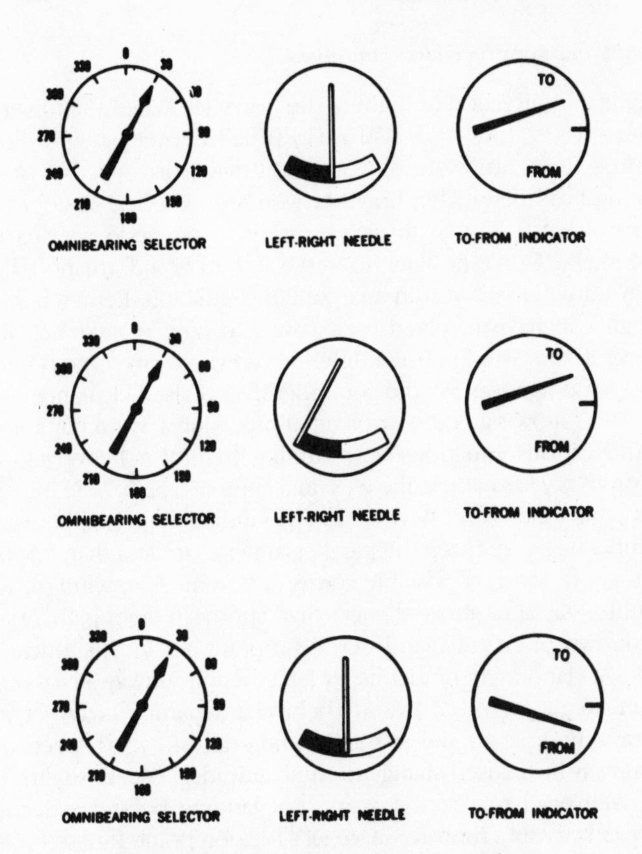

Figure V-5. VOR instrument examples.

VORTAC/VHF omnidirectional range/tactical air navigation—A navigation aid providing VOR azimuth, TACAN azimuth, and TACAN distance measuring equipment (DME) at one site.

vortex—(*See* WAKE TURBULENCE.)

vortex avoidance procedures—

1. General. Under certain conditions, airport traffic controllers apply procedures for separating other aircraft from heavy jets. They will also provide VFR aircraft, which in the tower's opinion may be adversely affected by potential wake turbulence, with the position,

vortex avoidance procedures *continued*

altitude and direction of flight of the heavy jet. When the tower controller advises "CAUTION WAKE TURBULENCE," etc., he is following his procedures and warning you that it may exist. Whether or not a warning has been given, however, you are expected to adjust your operation and flight path as necessary to preclude serious wake encounters. Don't hesitate to request further information if you believe it will assist you in analyzing the situation. Remember, even though you have received a clearance to land or take off, if you believe it safer to wait, use a different runway, or in some other way alter your operation, ask the controller for a revised clearance.

2. The following vortex avoidance procedures are recommended for the various situations: (a) landing behind a heavy jet, same runway—stay at or above the jet's final approach path, note his touchdown point, and land beyond it; (b) landing on a parallel runway behind a heavy jet when the parallel runways are less than 2,500 feet apart—note wind for possible vortex drift, request upwind runway if possible. Stay at or above the jet's final approach flight path, note his touchdown point, and land beyond a point abeam his touchdown point; (c) landing behind a heavy jet, crossing runway—cross above the jet's flight path; (d) landing behind a departing heavy jet, same runway—note jet's rotation point (where jet's forward speed down the runway has caused change to climb attitude, nose up slightly) and land well prior to rotation point; (e) landing behind a departing heavy jet, crossing runway—note jet's rotation point. If past the intersection, land prior to the intersection. If jet rotates prior to the intersection avoid flight below the jet's flight path. Abandon the approach unless landing is assured well before reaching the intersection; (f) departing behind a heavy jet—the tower will withhold clearances for 2 minutes for takeoffs on the same runway, a parallel runway separated by less than 2,500 feet, and any other situation where inflight crossing courses are evident. Pilots should note the jet's rotation point, and climb above the jet's climb path until turning clear of his wake. Avoid subsequent flight paths which will cross below and behind a heavy jet; (g) intersection takeoffs, same runway—towers will withhold take-off clearance for 3 minutes behind a large turbojet on the same runway. Be alert to adjacent heavy jet operations particu-

larly upwind of your runway. Avoid subsequent headings which will cross below a heavy jet's path; and (h) en route VFR—avoid flight below and behind a heavy jet's flight path. If you observe a heavy jet above you on the same track (same or opposite direction) adjust your position laterally, preferably upwind to the jet.

vortex characteristics—Trailing vortex wakes have certain characteristics. Being familiar with these characteristics will assist the pilot in visualizing the location of these vortex wakes.

1. Vortex generation starts with rotation for lift-off and ends when the wing unloads after touchdown. [*See* Figure V-6(a).]

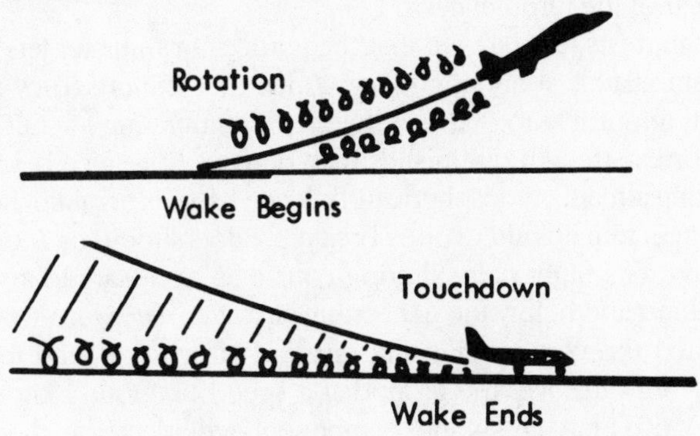

Figure V-6(a). Vortex generation wakes.

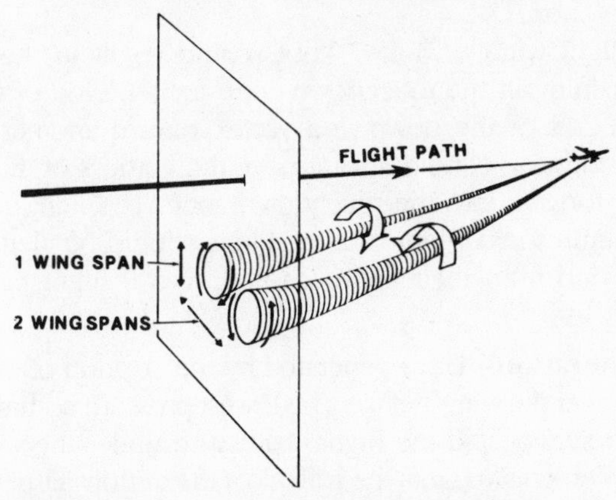

Figure V-6(b). Vortex flow field in aircraft wake.

vortex characteristics *continued*

2. The vortex circulation is outward, upward and around the wing tips when viewed from either ahead or behind the aircraft. Tests with large aircraft have shown that the vortex flow field, in a plane cutting through the wake at any point downstream, covers an area about 2 wing spans in width and one wing span in depth. The vorteces remain so spaced (about a wingspan apart) even drifting with the wind, at altitudes greater than a wing span from the ground. In view of this, if persistent vortex turbulence is encountered, a slight change of altitude and lateral position (preferably upwind) will provide a flight path clear of the turbulence.

3. Flight tests have shown that the vortices from heavy jets start to sink immediately at about 400 to 500 feet per minute. They tend to level off about 800 to 900 feet below the generating aircraft's flight path. Vortex strength diminishes with time and distance behind the generating aircraft. Atmospheric turbulence hastens breakup. Residual choppiness remains after vortex breakup. Pilots should fly at or above the heavy jet's flight path, altering course as necessary to avoid the area behind and below the generating aircraft. [*See* Figure V-6(b).]

4. When the vortices sink into ground effect they tend to move laterally outward over the ground at a speed of about 5 knots. [*See* Figure V-6(d).] A crosswind component will decrease the lateral movement of the upwind vortex and increase the movement of the downwind vortex.

Thus a light wind of 3 to 7 knots could result in the upwind vortex remaining in the touchdown zone for a period of time and hasten the drift of the downwind vortex toward another runway. Similarly, a tailwind condition can move the vortices of the preceding aircraft forward into the touchdown zone. The light quartering tailwind requires maximum caution. Pilots should be alert to large aircraft upwind from their approach and takeoff flight paths. [*See* Figure V-6(e).]

vortex generation—Lift is generated by the creation of a pressure differential over the wing surface. The lowest pressure occurs over the upper wing surface and the highest pressure under the wing. This pressure differential triggers the roll up of the airflow aft of the wing resulting in swirling air masses trailing downstream of the wing tips.

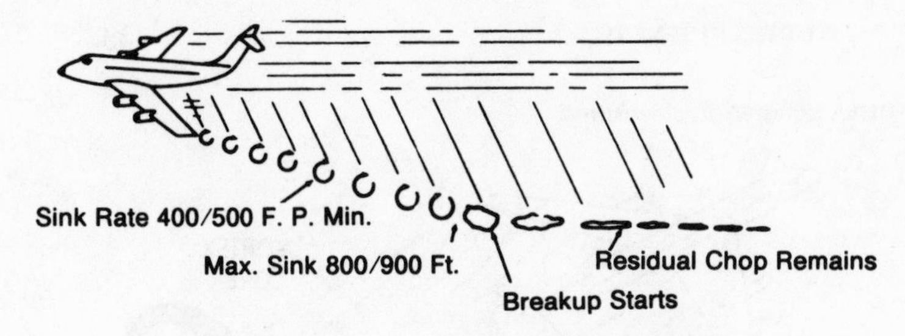

Figure V-6(c). Vortex strength diminishes.

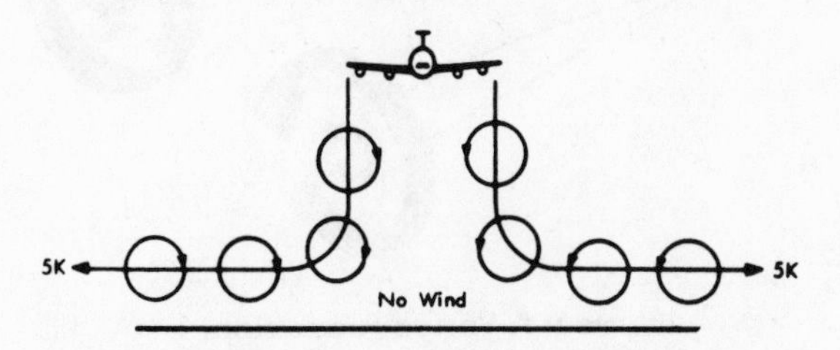

Figure V-6(d). Vortex movement in ground effect.

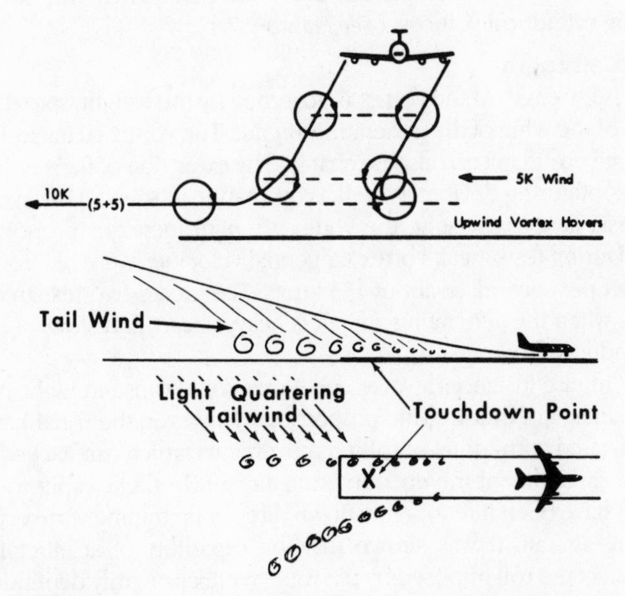

Figure V-6(e). Vortex movement in ground effect with cross wind and tail wind.

vortex generation *continued*

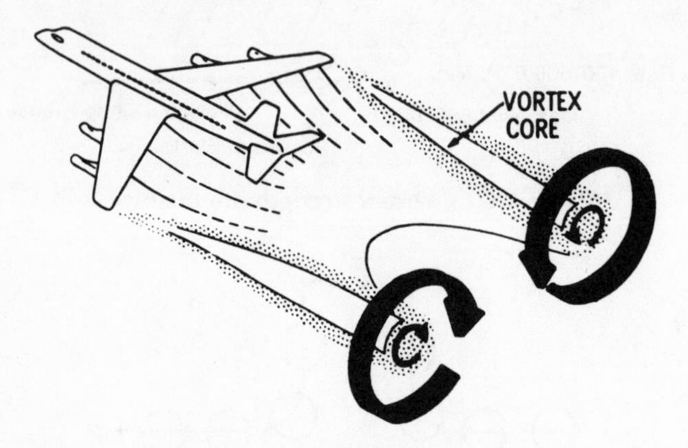

Figure V-7. Vortex generation.

After the roll up is completed, the wake consists of two counter rotating cylindrical vortices. (*See* Figure V-7.)

vortex strength—

1. The strength of the vortex is governed by the weight, speed, and shape of the wing of the generating aircraft. The vortex characteristics of any given aircraft can also be changed by extension of flaps or other wing configuring devices as well as by change in speed. However, as the basic factor is weight, the vortex strength increases proportionately. During tests, peak vortex tangential velocities were recorded at 224 feet per second, or about 133 knots. The greatest vortex strength occurs when the generating aircraft is heavy, clean, and slow.

2. Induced Roll

a. In rare instances a wake encounter could cause in-flight structural damage of catastrophic proportions. However, the usual hazard is associated with induced rolling movements which can exceed the rolling capability of the encountering aircraft. In flight experiments, aircraft have been intentionally flown directly up trailing vortex cores of large aircraft. It was shown that the capability of an aircraft to counteract the roll imposed by the wake vortex primarily depends on the wing span and counter responsiveness of the encountering aircraft.

b. Counter control is usually effective and induced roll minimal in cases where the wing span and ailerons of the encountering aircraft extend beyond the rotational flow field of the vortex. It is more difficult for aircraft with short wing span (relative to the generating aircraft) to counter the imposed roll induced by vortex flow. Pilots of short span aircraft, even of the high performance type, must be especially alert to vortex encounters. (*See* Figure V-8)

c. The wake of large aircraft requires the respect of all pilots.

vortices/wingtip vortices—Circular patterns of air created by the movement of an airfoil through the air when generating lift. As an airfoil moves through the atmosphere in sustained flight, an area of low pressure is created above it. The air flowing from the high pressure area to the low pressure area around and about the tips of the airfoil tends to roll up into two rapidly rotating vortices, conical in shape. These vortices are the most predominant parts of aircraft wake turbulence and their rotational force is dependent upon the wing loading, gross weight, and speed of the generating aircraft. The vortices from medium to heavy aircraft can be of extremely high velocity and hazardous to smaller aircraft. (*See* VORTEX AVOIDANCE PROCEDURES.)

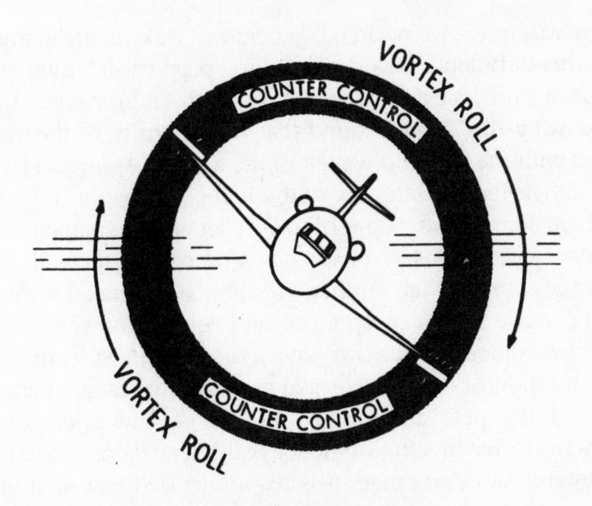

Figure V-8. Vortex roll.

very high frequency omnidirectional range station (VOR)—A ground-based electronic navigation aid transmitting very high frequency navigation signals, 360 degrees in azimuth, oriented from magnetic north. Used as the basis for navigation in the national airspace system. The VOR periodically identifies itself by Morse code and may have an additional voice identification feature. Voice features may be used by ATC or FSS for transmitting instructions/information to pilots.

VOT—A VOR receiver testing facility.

vrbl—Variable.

vsby—Visibility.

W in the phonetic alphabet is Whiskey (wiss-key).

W—A symbol for an air mass warmer than the surface over which it is moving.

wake turbulence—Every aircraft generates a wake while in flight. In the past this turbulence was attributed to "prop wash." Later studies found this disturbance to be a pair of counterrotating vortices trailing from the wing tips. It was found that the intensity of the vortices increased with the size and weight of the aircraft. As aircraft became larger and heavier, the intensity of the vortices began to pose problems for smaller aircraft. Some of today's jet aircraft, particularly the new jumbo jets, generate roll velocities exceeding the roll control capability of some aircraft. Further, turbulence generated within vortices can damage aircraft components and equipment when encountered at close range. For his own safety, the pilot must learn to envision the location of the vortex wake generated by large aircraft and adjust his flight path accordingly. During ground operations, jet engine blast (thrust stream turbulence) can cause damage and upsets if encountered at close range. It is recommended that light aircraft remain at least 600 feet behind a jet operating its engines at idle speed, 1,250 feet behind a taxiing jet, and 2,100 feet from a jet taking off. Engine thrust velocities generated by large jet aircraft during initial takeoff roll, and the drifting of the turbulence in relation to the crosswind component dictate the desirability of lighter aircraft awaiting takeoff to hold well back of the runway edge or taxiway hold line; also, aligning the aircraft to face the possible jet engine blast movement is desirable. The Federal Aviation Administration has established standards for location of taxiway hold lines at airports served by air carriers as follows: "Locate all taxiway holding lines such that the distance from the runway structural pavement edge to the taxiway holding line is at least equal to the greater of the following: (1) 100 feet or (2) 150 feet when heavy jets are expected to use the runway." (*Also see* VORTEX AVOIDANCE PROCEDURES.)

wake turbulence, operational problem areas—

1. A wake encounter may not necessarily be hazardous. It can be one or more jolts with varying intensity, depending on the direction of the encounter, distance from the generating aircraft, and point of vortex encounter. The probability of buffeting and induced roll increases when the encountering aircraft's heading is generally aligned with the vortex trail. Pilots should be particularly alert to calm wind conditions and situations when the vortices: (a) remain in the touchdown area; (b) drift downwind to a parallel runway; (c) sink into the take-off or landing path of a crossing runway; (d) sink into the traffic patterns for other airports; (e) sink into the flight path of VFR flights operating below.

2. Pilots should visualize the location of the vortex trail and use proper avoidance procedures in a mixed traffic environment. Avoid wake encounters below and behind the generating aircraft, especially at low altitude where even a momentary wake encounter could be hazardous. (*Also see* VORTEX AVOIDANCE PROCEDURES.)

wake turbulence separations—

1. Because of the possible effects of wake turbulence, controllers are required to apply no less than specified minimum separation for aircraft operating behind a heavy jet and, in certain instances, behind large nonheavy aircraft.

wake turbulence separations *continued*

a. Separation is applied to aircraft operating directly behind a heavy jet at the same altitude or less than 1,000 feet below:

i. Heavy jet behind heavy jet—4 miles.

ii. Small/large aircraft behind heavy jet—5 miles.

b. Also, separation, measured at the time the preceding aircraft is over the landing threshold, is provided to small aircraft:

i. Small aircraft landing behind heavy jet—6 miles.

ii. Small aircraft landing behind large aircraft—4 miles.

c. Additionally, appropriate time or distance intervals are provided to departing aircraft:

i. Two minutes or the appropriate 4- or 5-mile radar separation when takeoff behind a heavy jet will be: from the same threshold; on a crossing runway and projected flight paths will cross; from the threshold of a parallel runway when staggered ahead of that of the adjacent runway by less than 500 feet and when the runways are separated by less than 2,500 feet.

2. A 3-minute interval will be provided when a small aircraft will take off:

a. From an intersection on the same runway (same or opposite direction) behind a departing large aircraft,

b. In the opposite direction on the same runway behind a large aircraft takeoff or low/missed approach.

NOTE: This 3-minute interval may be waived upon specific pilot request.

3. A 3-minute interval will be provided for all aircraft taking off when the operations are as described in 2a and 2b above, the preceding aircraft is a heavy jet, and the operations are on either the same runway or parallel runways separated by less than 2,500 feet. Controllers may not reduce or waive this interval.

4. Pilots may request additional separation, i.e., 2 minutes instead of 4 or 5 miles for wake turbulence avoidance. This request should be made as soon as practical on ground control and at least before taxiing onto the runway.

warning areas—Airspace which may contain hazards to nonparticipating aircraft in international airspace. Warning areas are established beyond the 3-mile limit. Though the activities conducted within warning areas may be as hazardous as those in restricted areas, warning areas cannot be legally designated because they are over international waters. Penetration of warning areas during periods of activity may be hazardous to the aircraft and its occupants. Official descriptions of warning areas may be obtained on request to the Federal Aviation Administration, Washington, D.C.

wash—The disturbed air in the wake of an airplane, particularly behind its propeller.

water, flight over—(*See* DITCHING OVER WATER.)

way point, RNAV—A predetermined geographical position, used for route or instrument approach definition or progress reporting purposes, that is defined relative to a VORTAC station position. Two subsequently related way points define a route segment.

wdly—Widely.

wea—Weather.

weather advisories, in-flight—

1. In-flight weather information is available by calling any FAA/FSS facility within radio range. Selected FSS's broadcast current weather reports, in-flight advisories, PIREPs, RAREPs, and NOTAMs at 15 minutes past every hour. TWEB also can be received in the air. Monitor weather broadcasts routinely and do not hesitate to request specific information from FAA/FSS—En Route Flight Advisory Service (EFAS)—"Flight Watch" at 122.0 MHz.

2. The purpose of this service is to notify en route pilots of the possibility of encountering hazardous flying conditions which probably were not provided in the preflight briefings. Whether or not the condition described is potentially hazardous to a particular flight is for the pilot himself to evaluate on the basis of his own experience and the operational limits of his aircraft. (*Also see* SIGMET and AIRMET.)

weather advisory—In aviation forecast practice, an expression of hazardous weather conditions not predicted in the area forecast, as they affect the operation of air traffic and as prepared by the national weather service.

"Weather, A.M."—The "A.M. Weather" television program seen on 250 public stations nationwide. Produced by the Maryland Center for Public Broadcasting, the 15-minute presentation airs weekday mornings, providing complete national weather plus special reports for aviation, agricultural, marine and recreation audiences.

Aviation weather information presented by A.M. Weather includes IFR, VFR, and marginal VFR flying weather, current and 12-hour weather depiction charts, turbulence information, winds aloft reports, and 24-hour outlooks on Thursday and Friday.

weather briefing—

1. Consult your local flight service station (FSS), or Weather Service Office (WSO) for preflight weather briefing.

2. When requesting a preflight briefing, identify yourself as a pilot and provide the following: (a) Type of flight planned, e.g., VFR or IFR. (b) Aircraft number or pilot's name. (c) Aircraft type. (d) Departure Airport. (e) Route of flight. (f) Destination. (g) Flight altitude(s). (h) ETD and ETE.

3. You are urged to use the pilot's preflight checklist which is on the reverse of the flight plan form. The checklist is a reminder of items you should be aware of before beginning flight, as well as a flight log for your use if desired. (*See* VFR FLIGHT PLAN.)

weather channel—A source of general weather information 24 hours a day on a cable TV channel.

weather minimums—(*See* VFR FLIGHT WEATHER MINIMUMS IN UNCONTROLLED AIRSPACE; VFR FLIGHT WEATHER MINIMUMS IN CONTROLLED AIRSPACE.)

weather radar—The national weather service operates a 56-station network of weather radars. These stations are generally spaced in such a manner as to enable them to detect and identify the type and characteristics of most of the precipitation east of the continental divide. In addition, 2 radars of the DOD and 22 FAA radars augment the network in the conterminous U.S. [*See* map, Figure W-1(a).]

weather, key to—[*See* Figure W-1(b).]

weathervane—The tendency of an airplane on the ground or water to face into the wind, due to its effect on the vertical surfaces of the tail group.

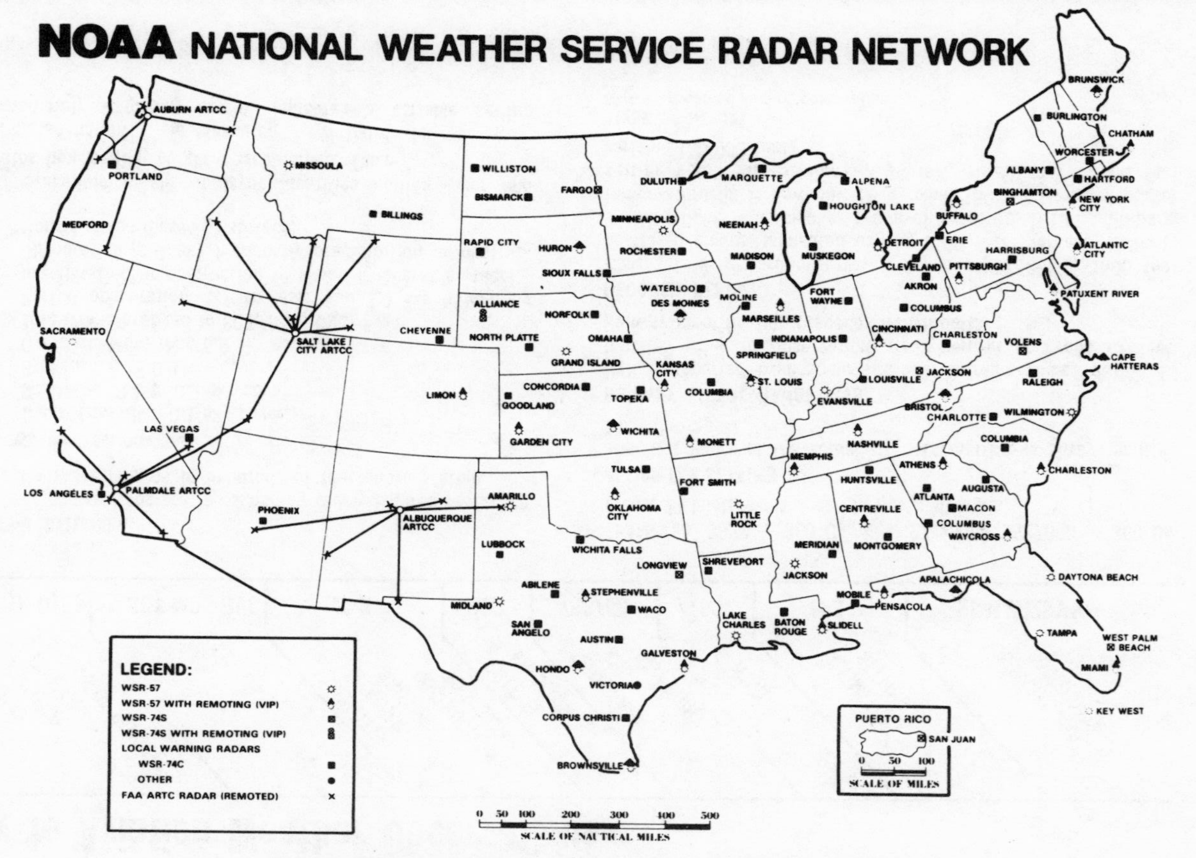

Figure W-1(a). NOAA National Weather Service Radar Network.

Key to Aviation Weather Observations

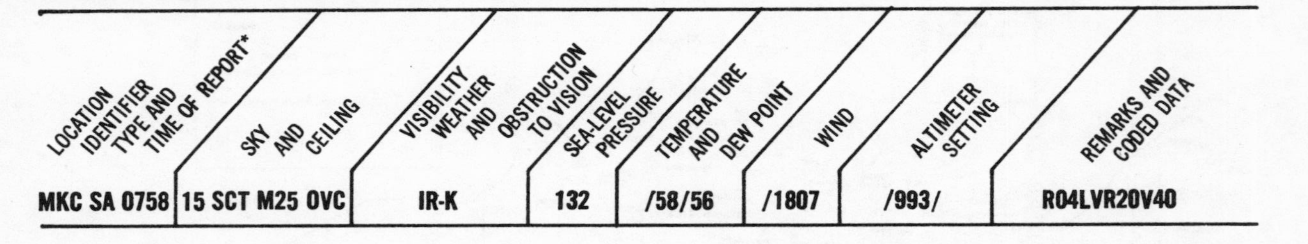

LOCATION IDENTIFIER TYPE AND TIME OF REPORT*	SKY AND CEILING	VISIBILITY WEATHER AND OBSTRUCTION TO VISION	SEA-LEVEL PRESSURE	TEMPERATURE AND DEW POINT	WIND	ALTIMETER SETTING	REMARKS AND CODED DATA
MKC SA 0758	15 SCT M25 OVC	IR-K	132	/58/56	/1807	/993/	R04LVR20V40

SKY AND CEILING

Sky cover contractions are in ascending order. Figures preceding contractions are heights in hundreds of feet above station.

Sky cover contractions are:

CLR Clear: Less than 0.1 sky cover.
SCT Scattered: 0.1 to 0.5 sky cover.
BKN Broken: 0.6 to 0.9 sky cover.
OVC Overcast: More than 0.9 sky cover.
— Thin (When prefixed to SCT, BKN, OVC)
—X Partial obscuration: 0.1 to less than 1.0 sky hidden by precipitation or obstruction to vision (bases at surface).
X Obscuration: 1.0 sky hidden by precipitation or obstruction to vision (bases at surface).

Letter preceding height of layer identifies ceiling layer and indicates how ceiling height was obtained. Thus:

E Estimated height **M** Measured **W** Indefinite
V Immediately following height, indicates a variable ceiling

EXAMPLES: 3627 = 360 Degrees; 27 knots; 3627G40 = 360 Degrees, 27 knots, peak speed in gusts 40 knots.

ALTIMETER SETTING

The first figure of the actual altimeter setting is always omitted from the report.

RUNWAY VISUAL RANGE (RVR)

RVR is reported from some stations. Extreme values during 10 minutes prior to observation are given in hundreds of feet. Runway identification precedes RVR report.

DECODED REPORT

Kansas City: Record observation taken at 0758GMT, 1500 feet scattered clouds, measured ceiling 2500 feet overcast, visibility 1 mile, light rain, smoke, sea-level pressure 1013.2 millibars, temperature 58 F. dewpoint 56°F, wind 180°, 7 knots, altimeter setting 29.93 inches. Runway 04 left, visual range 2000 feet variable to 4000 feet.

*TYPE OF REPORT
SA—a scheduled record observation

VISIBILITY

Reported in statute miles and fractions. (V=Variable)

WEATHER AND OBSTRUCTION TO VISION SYMBOLS

A	Hail	**SG**	Snow grains
BD	Blowing dust	**SP**	Snow pellets
BN	Blowing sand	**SW**	Snow showers
BS	Blowing snow	**T+**	Severe thunderstorm
D	Dust	**T**	Thunderstorm
F	Fog	**ZL**	Freezing drizzle
GF	Ground fog	**ZR**	Freezing rain
H	Haze	**K**	Smoke
IC	Ice crystals	**L**	Drizzle
IF	Ice fog	**R**	Rain
IP	Ice pellets	**RW**	Rain showers
IPW	Ice pellet showers	**S**	Snow

Precipitation intensities: —Light; (no sign) Moderate; + Heavy.

WIND

Direction in tens of degrees from true north, speed in knots. 0000 indicates calm. G indicates gusty. Peak speed of gusts follows G or Q when gusts or squall are reported. The contraction WSHFT followed by GMT time group in remarks indicates windshift and its time of occurrence.

SP—an unscheduled special observation indicating a significant change in one or more elements

RS—a scheduled record observation that also qualifies as a special observation.

All three types of observations (SA, SP, RS) are followed by a 24 hour-clock-time-group in GMT.

THE IMPORTANCE OF PIREP's

How can you help eliminate, or at least reduce en route weather surprises? The best way is—to fill-in the few-and-far-between surface weather observations—with YOUR PIREP's.

A PIREP gives valuable information that pilots actually experience in-flight in contrast to the data a weather observer measures or estimates on the ground.

Here's an example of encoded and decoded PIREP's showing the wealth of data a pilot can give and receive.

ENCODED

UA /OV FRR 275045 1745 FL330 /TP B727 /SK 185 BKN 220 /280 BKN 310 /TA -53 /WV 290120 /TB LGT-MDT-CAT ABV-310

DECODED

Pilot report, Front Royal VORTAC 275 radial 45nm, at 1745Z, flight level 33000; Boeing 727; cloud base 18500 broken, tops 22000, second layer 28000 broken tops 31000; air temperature minus 53 degrees Celsius; wind 290 degrees 120 knots; light to moderate clear air turbulence above 31000.

Figure W-1(b). Key to Aviation Weather Observations.

weathered-in—Forced to stay on the ground by bad weather.

weight and balance—Must be considered by every pilot before every flight; see owner's manual for particular aircraft's characteristics. NOTE: Some aircraft will exceed their gross weight limits if all seats are occupied and fuel tanks are full.

wheelbarrowing—Landing too fast in a tricycle-gear aircraft causing a porpoising effect.

wilco—Pilot's radio expression meaning I have received your message, understand it, and will comply with it.

wildlife refuge area—The unauthorized operation of aircraft at low altitudes over, or the unauthorized landing of aircraft on a wildlife refuge area is prohibited, except in the event of emergency. The Fish and Wildlife Service requests that pilots maintain a minimum altitude of 2,000 feet above the terrain of a wildlife refuge area.

wind direction indicator—A wind cone installed at the center of the segmented circle on an airport and used to indicate wind direction and velocity. The large end of the wind cone points into the wind.

wind direction indicators—At an airport without a control tower, may be a wind sock, tetrahedron or symbol in a segmented circle. (*See* Figures W-3, page 243; T-1, page 185; and S-1, page 168.)

wind velocities symbols on weather maps—(*See* Figure W-2.)

wind shear—A change in wind speed and/or wind direction in a short distance resulting in a tearing or shearing effect. It can exist in a horizontal or vertical direction and occasionally in both.

wind shift—Or wind shift line, an abrupt change in the direction or velocity, or both, of the wind. Usually associated with a front.

wind sock—A cloth sleeve, mounted aloft at an airport to use for estimating wind direction and velocity. (*See* Figure W-3.)

wind tee—An indicator for wind or traffic direction at an airport.

wing—An airfoil whose major function is to provide lift by the dynamic reaction of the mass of air swept downward.

wing bow—Used at wingtip to provide a rounded conformation. Sometimes used to denote the wing tip.

ing heavy—A condition of rigging in an airplane in which one ing tends to sink.

ing-over—A flight maneuver in which the airplane is alternately limbed and dived during a 180 degree turn.

ing root—The end of a wing which joins the fuselage, or the oppo-ite wing.

ing tip—The end of the wing farthest from the fuselage, or cabin.

VIP—Work in progress.

vk—Weak.

vords twice—Radio phraseology meaning:

1. As a request, "Communication is difficult. Please say every phrase wice."

2. As information, "Since communications are difficult, every phrase n this message will be spoken twice."

vorld aeronautical charts—(*See* AERONAUTICAL CHARTS.)

VS—Weather service.

vt—Weight.

vv—Wave.

VX—Weather

X

X in the phonetic alphabet is Xray (ecks-ray).

X-mark—On an airport runway it means runway is closed.

	Calm	Calm		44 - 49
	1 - 2	1 - 2		50 - 54
	3 - 8	3 - 7		55 - 60
	9 - 14	8 - 12		61 - 66
	15 - 20	13 - 17		67 - 71
	21 - 25	18 - 22		72 - 77
	26 - 31	23 - 27		78 - 83
	32 - 37	28 - 32		84 - 89
	38 - 43	33 - 37		119 - 123

Figure W-2. Wind velocities symbols on weather map

Y

Y in the phonetic alphabet is Yankee (yang-key).

yaw—To turn about the vertical axis. (An airplane is said to yaw as the nose turns without the accompanying appropriate bank.)

Z

Z in the phonetic alphabet is Zulu (zoo-loo).

Z—Greenwich mean time.

zones, control—(*See* CONTROL ZONE.)

zoom—To climb for a short time at an angle greater than the normal climbing angle, the airplane being carried upward by momentum.